Towards a Political Economy of Resource-dependent Regions

T0248669

This book advances our understanding of resource-dependent regions in developed economies in the 21st century. It explores how rural and small town places are working to find success in a new economy marked by demographic, economic, social, cultural, political, and environmental change. How are we to understand the changes and transformations working through communities and economies? Where are the trajectories of change leading these resource-dependent places and regions? Drawing upon examples from Canada, USA, UK, Australia, New Zealand, and the Nordic countries, these and other questions are explored and addressed by constructing a critical political economy framework of resource hinterland transition.

Towards a Political Economy of Resource-dependent Regions is a key resource for students and researchers in geography, rural and industrial sociology, economics, environmental studies, political science, regional studies, and planning, as well as policy makers, those in industry and the private sector, and local and regional development practitioners.

Greg Halseth is a Professor in the Geography Program at the University of Northern British Columbia, Canada, where he is also the Canada Research Chair in Rural and Small Town Studies and Co-Director of UNBC's Community Development Institute. His research examines rural and small town community development, and local and regional strategies for coping with social and economic change.

Laura Ryser is the Research Manager of the Rural and Small Town Studies Program at the University of Northern British Columbia, Canada. Her research interests include small town community change, institutional barriers to change, building resiliency to respond to restructuring trends, labour restructuring, and rural poverty.

Routledge Studies in Human Geography

This series provides a forum for innovative, vibrant, and critical debate within Human Geography. Titles will reflect the wealth of research which is taking place in this diverse and ever-expanding field. Contributions will be drawn from the main sub-disciplines and from innovative areas of work which have no particular sub-disciplinary allegiances.

For a full list of titles in this series, please visit www.routledge.com/series/SE0514

Towards a Political Economy of Resource-dependent Regions

Greg Halseth and Laura Ryser

Routledge
Taylor & Francis Group

LONDON AND NEW YORK

First published 2018 by Routledge

2 Park Square, Milton Park, Abingdon, Oxfordshire OX14 4RN
52 Vanderbilt Avenue, New York, NY 10017

Routledge is an imprint of the Taylor & Francis Group, an informa business

First issued in paperback 2019

Copyright© 2018 Greg Halseth and Laura Ryser

The right of Greg Halseth and Laura Ryser to be identified as authors of this work has been asserted by them in accordance with sections 77 and 78 of the Copyright, Designs and Patents Act 1988.

All rights reserved. No part of this book may be reprinted or reproduced or utilised in any form or by any electronic, mechanical, or other means, now known or hereafter invented, including photocopying and recording, or in any information storage or retrieval system, without permission in writing from the publishers.

Notice:
Product or corporate names may be trademarks or registered trademarks, and are used only for identification and explanation without intent to infringe.

British Library Cataloguing in Publication Data
A catalogue record for this book is available from the British Library

Library of Congress Cataloging in Publication Data
A catalog record for this book has been requested

ISBN: 978-0-415-78842-7 (hbk)
ISBN: 978-0-367-24529-0 (pbk)

Typeset in Times New Roman
by Taylor & Francis Books

For Our Fathers

Contents

Acknowledgements

This book is the product of years of living, working, and researching in resource-dependent towns and regions. As such, a key source of our learning comes from the communities who are themselves immersed in the changes associated with the late stages of industrial resource production. We are very grateful to the many people and many communities that have interacted with us and our research team over the years. We have learned so much from them in the ground-truthing of the theory and practice of social and economic restructuring.

One key way by which we engage with communities is through the work of UNBC's Community Development Institute (CDI). Formally established in 2004, the CDI has established a leading reputation in community-based research and has worked with communities on a host of issues ranging from economic, to social, to political, to cultural change. The foundation of the CDI's success and its engagement with communities is the work of long time CDI coordinator Don Manson. For close to 20 years, Don worked for the betterment of communities across the region. He pioneered research techniques and approaches that resulted in greater engagement and interaction between the university and communities. He helped change both people and communities for the better. We are very fortunate in that the CDI is now under the direction of Marleen Morris. Building from deep personal, family, and professional experiences, Marleen brings a heartfelt passion for helping communities address challenges and seize opportunities. As northern BC engages with ongoing rounds of social, political, and economic restructuring, Marleen and her team are working every day with people and communities to help them with the information and support they need to make better decisions about their own futures. Under Marleen's leadership, the CDI continues its proud tradition of "research having an impact; people making a difference".

In working to put this book together, we are also grateful to the various publishers who have granted permission to use previously published material. This includes:

Excerpts from *Investing in place* by Sean Markey, Greg Halseth, and Don Manson © University of British Columbia Press 2012, were reprinted with permission of the Publisher. All rights reserved by the Publisher.

Excerpts from *Flexible crossroads* by Roger Hayter © University of British Columbia Press 2000, were reprinted with permission of the Publisher. All rights reserved by the Publisher.

Excerpts from *A brief history of neoliberalism* by David Harvey © Oxford University Press 2005, were reprinted with permission of the Publisher. All rights reserved by the Publisher.

Excerpts from *The right to traditional resources and development program* by Steinar Pedersen © The Community Development Institute at UNBC 2015, were reprinted with permission of the Publisher. All rights reserved by the Publisher.

Excerpts were reprinted with permission of the Publisher from *Social transformation in rural Canada* edited by John Parkins and Maureen Reed © University of British Columbia Press 2012. All rights reserved by the Publisher.

Excerpts were reprinted with permission of the Publisher from *Transformation of resource towns and peripheries: Political economy perspectives* edited Greg Halseth © Routledge 2017. All rights reserved by the Publisher.

Excerpts were reprinted with permission of the Publisher from *The structure and dynamics of rural territories: Geographic perspectives* edited by Doug Ramsey and Christopher Bryant © Brandon University, Rural Development Institute 2004. All rights reserved by the Publisher.

Excerpts were reprinted by permission of Taylor & Francis LLC (http://www.tandfonline.com) from L. Ryser and G. Halseth (2016) "Opportunities and challenges to address poverty in rural regions: A case study from Northern BC", *Journal of Poverty* 21(2): 1–22. DOI: http://dx.doi.org/10.1080/10875549.2016.1141386. All rights reserved by the Publisher.

Excerpts were reprinted by permission of the Publisher from L. Ryser and G. Halseth (2014) "On the edge in rural Canada: The changing capacity and role of the voluntary sector", *Canadian Journal of Nonprofit and Social Economy Research* 5(1): 41–56. All rights reserved by the Publisher.

Excerpts were reprinted by permission of the Publisher from M. Porter and M. Kramer (2011) Creating shared value: How to reinvent new capitalism, and unleash a new wave of innovation and growth. *Harvard Business Review*, January / February: 62–77. All rights reserved by the Publisher.

Excerpts were reprinted by permission of the Publisher from G. Keenan (2015) "Tax shift: Companies dump burden of taxes on squeezed municipalities", *Globe and Mail* July 24. Available online at: http://www.theglobeandmail.com/report-on-business/tax-shift-companies-dump-burden-of-taxes-on-squeezed-municipalities/article25670719. Accessed February 2, 2016. All rights reserved by the Publisher.

Excerpts were reprinted from the article by S. Markey, L. Ryser and G. Halseth (2015) "'We're in this all together': Community impacts of long-distance labour commuting", *Rural Society* 24 (2): 131–153. Available online at: http://www.tandfonline.com/doi/full/10.1080/10371656.2015.1060717. Accessed February 2, 2016. Used by permission of the Publisher. All rights reserved by the Publisher.

Today, it is very rare that any researcher in social science works 'alone'. We have had the privilege to work with a great many colleagues over the years. They have provided stimulation and challenge as ideas around transformation were debated. We wish to recognize several sets of colleagues in particular. The first involves those here at UNBC. Across the university, the administration, staff, and faculty are committed to the betterment of northern British Columbia and advancing an understanding of the changes working through this region. We are grateful to so many who supported our research. In particular, we would like to acknowledge from the Geography program Gail Fondahl, Neil Hanlon, Catherine Nolin, Michelle Keen, Zoë Meletis, Brian Menounos, Roger Wheate, and Ellen Petticrew. From the Anthropology program Angèle Smith; from the Political Science program Gary Wilson, Michael Murphy, John Young, and Tracy Summerville; from the Social Work program Glen Schmidt, Si Transken, and Dawn Hemmingway; from First Nations Studies Ross Hoffman and Antonia Mills; from Environmental Studies Annie Booth; from Psychology Cindy Hardy; from Health Sciences Henry Harder, Chris Buse, and Margot Parkes; and from Ecosystem Science and Management Mike Gillingham, Hugues Massicotte, Catherine Parker, Art Fredeen, Scott Green, Kathy Lewis, Chris Johnson, Mark Shrimpton, Nikolaus Gantner, and our friend the late Lito Arocena. A special note is reserved for Kevin Hall. As founding chair of UNBC's Geography Program, Kevin provided selfless and valued guidance and mentorship that is both treasured and appreciated.

A second set includes colleagues we have come to know through the Canadian Rural Revitalization Foundation and its various research projects on the New Rural Economy. Led by Bill Reimer of Concordia University, this working group represented one of the most exciting opportunities in Canadian rural research. Included are: Bill Reimer (now retired from Concordia University), Tom Beckley (University of New Brunswick), David Bruce (Mount Allison University), Bruno Jean (Université du Québec à Rimouski), Patrice LeBlanc (Université du Québec en Abitibi-Témiscamingue), Doug Ramsey (Brandon University), Ellen Wall (now retired from University of Guelph), the late Derek Wilkinson (Laurentian University), Ivan Emke (Sir Wilfred Grenfell College, Memorial University of Newfoundland), Diane Martz (University of Saskatchewan), Omer Chouinard (Université de Moncton), Anna Woodrow (Concordia University), Steve Plante (Université du Québec à Rimouski), Dianne Looker (Acadia University), Peter Apedaile (now retired from the University of Alberta), and Ray Bollman (Statistics Canada).

Another valued group of Canadian colleagues include fellow researchers interested in resource development and non-metropolitan Canada. They include: Chris Bryant (Université du Montréal); Sarah Dorow, John Parkins, and Lars Hallström (University of Alberta); Hugh Gayler (Brock University); Ryan Gibson and David Douglas (University of Guelph); Alison Gill, Roger Hayter, and John Pierce (Simon Fraser University); Heather Hall (University of Waterloo); Lois Jackson (Dalhousie University); Rosemary Ommer and Aleck Ostry (University of Victoria); Nicole Power, Maura Hanrahan, Rob Greenwood, Sharon Roseman,

Keith Storey, Kelly Vodden, and Barb Neis (Memorial University of Newfoundland); Maureen Reed (University of Saskatchewan); the late Kent Sedgwick (College of New Caledonia); Mark Skinner (Trent University); Deatra Walsh (Nunavut); and Alun Joseph (retired from the University of Guelph).

A third group from whom we have learned a great deal is our international colleagues engaged in various aspects of rural research and development studies. Some of these colleagues have contributed to recent edited books such as *The next rural economies*, and *Transformation of resource towns and peripheries*. We would like to recognize and acknowledge Neil Argent (University of New England, Australia); Mary Cawley (National University of Ireland in Galway); Jean-Paul Charvet (Université of Paris Ouest Nanterre La Défense); Deborah Che, Owen Furuseth (University of North Carolina at Charlotte); Maija Halonen (University of Eastern Finland); Lisa Harrington (Kansas State University); Juha Kotilainen (University of Eastern Finland); Masatoshi Ouchi (Meiji University); David Storey (University of Worcester); Nigel Walford (Kingston University); Michael Woods (Aberystwyth University); Markku Tykkyläinen (University of Eastern Finland); Eero Vatanen (University of Eastern Finland); Sean Connelly and Etienne Nel (University of Otago); Dieter Müller (Umeå University); Mathew Tonts (University of Western Australia); Kieran Walsh (National University of Ireland – Galway); and Rachael Winterton (La Trobe University).

Another valued group of international colleagues include those second home researchers in Finland who are committed to better understanding the dynamic social and environmental impacts of our increasingly mobile society. This initiative is led by Mia Vepsäläinen, Kati Pitkänen, and Mervi Hiltunen who all came together at the University of Eastern Finland. They have gathered a remarkable group of colleagues including Seija Tuulentie and Asta Kietavainen at the Finnish Forest Research Institute, Antti Rehunen, Eeva Furman, and Leena Kopperoinen at the Finnish Environment Institute, Antti Honkanen and Olga Lipkina at the University of Eastern Finland, and the leading second home expert Michael Hall at the University of Canterbury in New Zealand.

We also wish to recognize our colleagues within the University of the Arctic's Thematic Network on Local and Regional Development in the North. This includes: Tor Gjertsen (founder and for ten years the lead for the Thematic Network) at UiT The Arctic University of Norway's Alta campus; Nils Aarsæther, Eva Schjetne, and Valeria Gjertsen also at UiT The Arctic University of Norway; Galina Knyazeva at Syktyvkar State University (Russia); Natalia Okhlopkova (former Director of the Finance and Economics Institute of North-Eastern Federal University); Oksana Romanova (current leader of the Thematic Network) at the North-Eastern Federal University (Republic of Sakha, Russia); Vyasheslav Shadrin (Chief of Council of Yukaghir Elders) who is a Research Fellow at The Institute for Humanities Research and Indigenous Studies of the North (IHRISN) Siberian Branch of the Russian Academy of Sciences; and Andra Aldea-Löppönen at the University of Oulu.

No listing of colleagues would be complete without recognizing our collaboration with Sean Markey from Simon Fraser University. Over the years and through many projects and adventures, over many miles and in many small communities, we could think of no better colleague and friend than Sean. Fortune smiled on our research when Simon Fraser University's John Pierce brought Sean on as a PhD student and then allowed him to work with us through his post-doctoral research.

Additional resources and time for the research and writing that underlies this book comes from the Federal Government's Canada Research Chair program. Beginning in 2002, Greg Halseth was awarded a Canada Research Chair tier two position, and in 2011 he was awarded a Canada Research Chair tier one position. This support is gratefully acknowledged as it allows for Canadian researchers to advance knowledge and understanding in support of their areas of expertise.

Our work has been tirelessly supported by the students and staff who over the years have formed the Rural and Small Town Studies team at UNBC. In particular, Lana Sullivan played a foundational role in establishing the team at UNBC and working on its early projects. Much of what we have achieved is based on the solid contributions and vision she brought. Other valued members of our team include Rosemary Raygada Watanabe, Julia Good (nee Schwamborn), and Kyle Kusch.

A great deal of the background research that contributed to our understanding of the changing political economy of resource-based communities was assisted by the graduate and undergraduate students of UNBC with whom we have had the pleasure to work with over the years. These students include: Alex Martin, Allison Matte, Amy Gondak, Anisa Zehtab-Martin, Anne Hogan, Ashley Kearns, Brian Stauffer, Carla Seguin, Catherine Fraser, Chelan Zirul, Christine Creyke, Colin McLeod, Courtney LeBourdais, Eric Kopetski, Gretchen Hernandez, Jennifer Crain, Jennifer Herkes, Jenny Lo, Jessica Raynor, Joanne Doddridge, Kelly Geisbrecht, Laura Van de Keere, Lila Bonnardel, Liz O'Connor, Marc Steynen, Melinda Worfolk, Melissa Zacharatos, Michelle White, Mollie Cudmore, Nora King, Onkar Buttar, Pam Tobin, Paul Pan, Priscilla Johnson, Rachael Clasby, Rebecca Goodenough, Rosalynd Curry, Shiloh Durkee, Tobi Araki, and Virginia Pow. Our recent team includes Gerald Pinchbeck, Daniel Bell, Alika Rajput, Alishia Lindsey, Kourtney Chingee, Devon Roy, Erin MacQuarrie, Danielle Patterson, Marli Bodhi, and Jessica Blewett.

We would also like to thank the publishing team at Routledge. Faye Leerink was our commissioning editor and has been a generous and patient supporter of the project. Pris Corbett was our editorial assistant and was a pleasure to work with. We would also like to thank Sarah M. Hall for her expert assistance in copyediting the manuscript. We are sure that readers will appreciate the fine index expertly crafted by Kathy Plett.

Perhaps most importantly, we draw our strength from our families. Thank you will never be a large enough expression!

Greg Halseth and Laura Ryser
May 2017

Part I
Introduction

Introduction

This book is interested in advancing an understanding of resource-dependent regions in developed economies at the opening of the 21st century. Resource production in remote, rural, and small town locations is certainly not new, and while change has always been a characteristic of these regions, the more rapid pace of transition and change for these regions and economies through the contemporary global economy has been well underway for more than 30 years. How are we to understand these regions? How are we to understand the changes and transformations working their way through communities and economies? What elements in these changes reflect emergence, what elements reflect transition, and what elements reflect continuity with the past? Where are the trajectories of change leading these remote, rural, and small town resource-dependent regions? It is to these questions that this book is addressed.

This introductory part of the book includes two chapters. Chapter 1 sets out our purpose and focus. Remote, rural, and small town resource-dependent regions in developed economies are continuing to change and transform. These changes and transformations are affecting the social, demographic, community, economic, and cultural structure of these places. To provide an 'umbrella' large enough to shelter an explanation for these varied issues, the book embarks on a search for a political economy of these regions. To ground and introduce that search, Chapter 2 provides thumbnail sketches of the intellectual traditions from which this search originates. As social geographers, there is first an introduction to the notion of geography as a perspective for understanding and making sense of the complex dynamic that is the world around us. The chapter then introduces the sub-discipline of social geography, followed by an introduction to the topic and framework of political economy. With an interest in the social, economic, and political dynamics of change, this framework provides the breadth to incorporate the many elements of change being experienced in resource-dependent regions.

By necessity, this book is written at a general level. The story is told with broad brush strokes and is highlighted with illustrations from many remote, rural, and small town resource-dependent regions from developed economies around the world. As will be described later, each place and region has a unique history and context, such that the general aspects of theory and explanation never fully replicate 'on the ground'. It is to the general level of explanation that we are interested, all the while recognizing that readers will situate and interpret the arguments we make through the lens of the places with which they are most familiar.

1 Introduction

Introduction

This book is about creating a framework for understanding the broader and complex context of resource-dependent places and regions within developed economies. It is also about organizing a wide body of theoretical ideas within that framework. The goal is to provide a guide to the forces of change so that community residents and leaders, policy makers, students, and researchers can better understand, and thus make better choices, about how to support more resilient communities and more sustainable economies.

Accelerating change has been a defining attribute of rural and small town places around the world. Traditional responses to these accelerating processes of change have too often only exacerbated the negative impacts of population loss and environmental degradation. As the reader will have noted already, our focus is with developed economies; and in these settings resource-dependent rural and small town places continue to struggle to find success in a new economy marked by demographic, economic, social, cultural, political, and environmental change. Older resource producing economies are transitioning within a 'hyper-connected' and increasingly 'commodified' global economy. We need to find more appropriate frameworks for understanding what is taking place, so that we can better shape how to respond.

This book will address the need to better understand these changes by constructing a critical political economy framework for resource hinterland transition within developed economies at the start of the 21st century. The purpose is to explore: 1) how local, regional, and global inter-linkages are affecting rural and remote resource-dependent communities; 2) the processes affecting the local capacity to respond to these changes; and 3) the strategic options available in light of these changes.

We chose our focus on developed economies for a number of reasons. To start, this is the geographic context of where we 'live' – both physically and as researchers. It is the economic and community landscape that we understand best and where we most likely are to encounter the fewest 'lost in translation' problems. Second, these landscapes have been experiencing accelerating change since the early 1980s and thus attention to supporting new community development pathways and investments is not only a concern, but also an urgent subject to which our engaged scholarship can contribute. This does not mean that we ignore issues in other parts of the world – variously labelled over time as concerning 'developing economies', the 'global south' (which makes little sense, since Australia and New Zealand are quite clearly south of the equator), or with acronyms such as the 'BRIC' states (Brazil, Russia, India, China). The value and problems with these labels we leave to the rich literature from development studies and other scholars. We recognize the interconnectivity of the global economy, and that changes in one region impact other regions. We recognize that economic dominance has severely impacted and limited developing regions and that industrial resource exploitation in those developing regions has impacts on the competiveness of older industries in developed economies. While there is a large literature in development studies, we focus on the context we know best.

With our focus, this book situates a collected body of knowledge and research respecting remote, rural, and small town resource-dependent regions within a larger framework for conceptualizing change. New approaches to understanding rural change, and the deployment of development strategies in the new rural economy, focus on economic and community development that are defined by both external and internal factors. For increasingly complex and connected economies, current pressures and future developments will require the construction of an integrated critical political economy framework of rural and small town transition. Through this book, we explore emerging questions around these internal and external factors to inform a richer understanding of the dynamics of change and transition.

Resource-dependent places

As noted, we are interested in creating a broad portrait of the dynamics and trajectories of change looking forward into the first half of the 21st century. With this intent, we are interested in rural and small town settlements, where the economies are dominated by one or more resource industry activities, and where those industries are organized around the extraction and export of the raw commodity itself, without engaging in any significant upstream processing or manufacturing. As we shall see later in the book, this economic orientation leaves the local economy highly vulnerable to the demands and prices set for natural resource commodities in the global marketplace. It also leaves them vulnerable to the strategies of international capital and its investment choices around the world.

Existing definitional frameworks of resource-dependent places and regions might talk about how a certain percentage of the labour force is employed in a single economic sector, in a single industry, or even with a single resource company. Other definitional frameworks might focus upon the particular type of resource commodity being produced. These different definitional frameworks all have use and value depending upon the purpose to which they are to be put. However, statistical or numerical 'breakpoints' in data or statistics which are then used to categorize places or regions as being 'in' or 'out' of one or another definitional category too often miss the nuances that are most critical in constructing an understanding of the character of places and their economies.

In general, resource-dependent regions tend to be characterized by a relative remoteness from the political and economic decision-making centres of their respective states. They also tend to be far from metropolitan cores. In terms of settlement structure, these regions tend to be comprised of numbers of small places, each with limited population and economic bases. Their economies are focused upon some aspect of the resource exploitation activities needed to prepare the commodity for export. Commensurate with their small populations, these places also tend to have relatively limited business and service sectors outside of those needed for household reproduction or for support of the local / regional resource industry. In part due to their remoteness and in part due to the way few of these settlements have shown marked growth over the past couple of decades, it is common for these places to be lagging with respect to investments in new quality-of-life services as well as the infrastructure to support new information and communications technologies.

Resource extraction activities within developed economies have a long history. Our focus is upon the post-Second World War period. This is when an industrial resource-development model took hold across most western developed economies. Additionally, our focus is on natural resource commodities derived from mining, forestry, fishing, and energy production. While mentioned in some discussions, we do not focus significant attention upon agriculture. Much like the debate about whether to focus on developed and developing economies together, the rural studies literature has for generations been synonymous with agricultural studies. As a result, there are, and continue to be, high quality and extensive literatures focused specifically upon agricultural issues and the social, political, and economic restructuring occurring within agricultural regions. It is our hope that this book will complement that work by telling the story of other resource sector regions and communities.

Restructuring

Research on local, regional, and global inter-linkages includes economic (drawing heavily upon the well-developed resource economy literatures) and political (drawing upon the emerging literatures connected with Neoliberalism, as well as more established 'new regional geography' literatures) restructuring. While the notion of 'restructuring' emerged through theoretical debate around the changing nature of capitalism, we focus on the elements, components, and foundations now underscoring rural change. In some cases, these activities work themselves through the different mechanisms and structures governing rural change. Important foci of attention include the processes of industrial consolidation, the adoption of workplace flexibility and labour shedding technologies, studies of corporate strategies to improve efficiencies, the impacts of diseconomies of scale due to rising energy costs and longer distances to transport raw materials, the impacts of currency fluctuations, commodity prices, land prices, and the costs of raw materials and supplies, and the uncertainty surrounding international trade agreements (Prudham, 2008). Political restructuring has led to the closure of local services, the adoption of 'enabling' policy approaches by senior governments, and the application of private sector or market measures of effectiveness and efficiency in public programming and supports (Young and Matthews, 2007). These measures and approaches are not well suited to the community context of remote, rural, and small town places, and they often diminish the capacity of these places to successfully adjust to change (Troughton, 2005).

Community development, place-based development, and new regionalism

Understanding and evaluating the processes affecting community capacity and local responsiveness requires attention to governance and decision making (with attention to literatures on community governance and the structures of power and conflict), as well as community capacity building (with attention to literatures on community development, community economic development, human and social capital, leadership, services, and the voluntary sector). The following discussion provides thumbnail sketches of community development, place-based development, and new regionalism. While they will recur later in the book, it is important to set them out early to help ground our argument.

Community development

In resource-dependent regions, there is an inter-connectedness of human capital, social capital, natural capital, and community capacity in the creation of place-based resilience to economic transition. However, a significant challenge for remote, rural, and small town places is that human capital development is increasingly a local responsibility (Joshi et al., 2000). This is problematic as such places, due to their small size, are especially sensitive to the 'leakage' of human capital (Dewees et al., 2003). Increasing labour-force participation by women (Little, 2002), the emergence of the knowledge economy, new forms of mobile labour activity, and the shifting role of immigration all highlight the need for broader attention to building and maintaining human capital in non-metropolitan regions.

At its simplest, human capital refers to the skills, knowledge, educational background, etc. of individuals. Human capital can be built with both experience and investment in training or education. Assessments and evaluations of human capital can be undertaken at the individual level; however, it can also be undertaken at the place or regional level. A more complex under-standing of human capital emerges when we link it with the economy and with concepts such as change. To start first with the economy, assessments of individual or collective human capital often focus upon the fit or lack of fit to the current economy. Do the job skills match those needed in local industry or businesses? Usually they do, since the local labour market and the local economy often grow and adjust together. But what about rapid change? New businesses or industries that require different skills sets may encounter a severe lack of fit with existing local human capital assets and as a result may need to turn, at least in the short term, to mobile labour in order to address skill or knowledge shortages. At a more general level, this also raises the question of whether the body of human capital is skilled in adaptation and whether it has the capacity to take advantage of new opportunities or adjust to rapidly changing circumstances. Literacy levels and ongoing measures of life-long learning are indicators of the resilience capacity that is needed within human capital.

Theoretical debate has used the concepts of social cohesion and social capital to understand how communal responses of shifting human capital work themselves out. Social cohesion is about the processes of relationships and interactions that build toward cohesive networks to support interaction. The products of these processes are the focus of social capital – the trust and bonds that develop from interactions in networks. While social cohesion can support opportunities for interaction through 'well-worn pathways' of collective action (Shortall, 2008), a lack of communication or the reorganization of social networks due to mobility and migration can limit its effectiveness. Although the literature is replete with interpretations, social capital can be operationalized in rural development research to refer to networks that produce the trust, reciprocity, and cooperation needed to mobilize in more locally appropriate and sustainable ways a host of social and economic resources (Kusakabe, 2012). Rural regions with strong internal and external networks are more likely to grow and support mechanisms for learning and innovation (Crowe, 2006). Bridging social capital (external links) is especially important in helping rural places acquire new forms of resources, information, advice, and training to support success in the emergent rural economy (Cawley and Nguyen, 2008).

Factors that support or impede social cohesion and social capital processes include the quality of networks and the dynamics of local actor cooperation. They also include the relative distribution of responsibilities, resources, or influences, and the openness of partnerships to creatively address local needs (Derkzen et al., 2008). Given the importance of local leadership (Bryant, 2010) in community development, we must pay attention to the increasing role played by the voluntary and non-profit sectors in rural and small town places, and the changing leadership (and the leadership mentoring) roles of local government and resource sector companies.

In numerous studies, processes of social cohesion and social capital have been shown as important in rural and regional development. They have been assessed as factors in development programming (Nardone et al., 2010), and they have been identified as important in supporting processes of innovation at local scales (Jakobsen and Lorentzen, 2015). Caution is needed, however, when interpreting concepts such as social capital and social cohesion. Too often, writers want to attach normative values such that social capital and social cohesion are interpreted as 'good' for community development. We caution against such normative valuations. In social capital and social cohesion, the issue is rather how they are deployed. If used to support participation, community engagement, access to information, etc., then the value might be clear. If, however, social capital and social cohesion are so strongly developed that communities close themselves off to new people and new ideas, then they may be counterproductive to community development.

Place-based development

Research on remote, rural, and small town restructuring has identified 'place' as a fundamental ingredient within the rural economy, where locale matters more than sector and competitive advantage more than comparative advantage. These arguments build from a recognition that within the global economy technology is shrinking the relevance of distance. As 'space' has become less of an issue in economic development, it is found that 'place' is becoming more important. Places are where the decisions, policies, and imperatives of regional, provincial, national, and international policies, investments, decisions, and activities act out.

Place-based economies recognize that the unique attributes and assets of individual places now determine their attractiveness for particular types of activities and investments (Woods, 2010), are more flexible and responsive to local needs, use local knowledge to enhance the appropriateness of initiatives (Courtney et al., 2006), and are more likely to focus upon the retention of benefits. To respond to the myriad of community, social, economic, cultural, and environmental changes, communities need to realistically consider and enumerate their place-based assets. Then they need to creatively re-imagine these assets, and re-bundle them to create new competitive advantages for their place and region. How they do this needs to be linked to their aspirations as a community *vis-à-vis* the type of place, society, and economy they wish to create.

Most place-based initiatives focus on product development (agriculture, tourism, etc.) and are undertaken to add value and gain market share or recognition. They may also support the creation of new place images or reinforce community cohesion by protecting or promoting local customs, languages, costumes, foods, crafts, festivals, traditions, and ways of life. Collectively, this is described as rural 'place-making' (Garrod et al., 2006). The fragmented nature of place-based initiatives may mean there are insufficient resources or limited capacity to address all development, planning, and implementation issues. As a result, there are concomitant needs for greater local and regional collaboration as well as supportive top-down policy and implementation support from senior governments (Connors, 2010).

Attention to the context and structure of places can also help address a range of challenges. Centralized public policy, especially with regard to matters of regional and local development, increasingly struggle with the pace of social and economic change and are subjected to often withering public scrutiny about the efficiency and appropriateness of policy application. As Bradford (2005) has argued, it is in places that today's major public policy issues play out, and those issues tend to be very resistant to sectoral interventions – they require coordination in place. Place-based policy approaches, therefore, can build from a shared foundation of knowledge for decision making, can develop synergies and a bottom-up generation of ideas and directions, can build the communication and collaboration needed to support policy development and implementation, and can provide the flexibility to develop closer links between a wide range of strategic planning processes and partners. The success of place-based policy initiatives, however, still requires the active involvement of senior government and the private sector. In some cases, this is because senior government retains control of critical policy and fiscal levers; in other cases, it is to support the "mechanisms which build on local capabilities and promote innovative ideas through the interaction of local and general knowledge and of endogenous and exogenous actors" (Barca et al., 2012 p. 149).

New regionalism

While the preceding 30 years have seen a rise in local approaches to development in concert with senior government policy withdrawal from direct engagement with rural and small town places and regions, the 'new regionalist' literature has emerged to support a return to regional approaches (Halseth et al., 2010b). Understandings of competitiveness and conceptualizations of a new rural economy have underlined the importance of strategic investments and supportive public policy in helping localist initiatives to succeed (Markey et al., 2008a). The comparative advantage approaches that drove much of post-Second World War development of resource commodity production have been disrupted by the entrance of many low-cost production regions with their lower wages, fewer and more permissive regulations, and lower tax regimes. This was coupled with the replacement of a Keynesian policy approach that saw senior government play an active, purposeful, and guiding role in regional development with what is now more generally called a Neoliberal policy approach (Sullivan et al., 2014). This emergent approach has led to a withdrawal of the state from purposeful engagement on the grounds that the 'market mechanisms' of private enterprise are more efficient and faster reacting.

Attention to these factors has supported experiments in regional governance as bottom-up approaches are re-scaled to facilitate multi-level development planning and implementation activities that are beyond the capacity of a single small community. Regional governance includes not only structural dynamics, but also issues of responsibility, reciprocity, performance, and norms of collaboration and inclusiveness (Ingamells, 2007). Halseth et al.

(2010b) add that regional governance will only be effective if accompanied by sufficient resources and allowed time to develop mature leadership, trust, and structure. It is important to understand how the changing structures of government and governance are responding to local and extra-local impulses of social and economic stress to avoid places competing against one another in a 'race to the bottom' approach to development.

Outline of book

The book is organized into four parts: 1) Introduction; 2) Organizing structures that are shaping the political economy of resource-dependent regions; 3) Mobilizing action through institutions; and 4) Change, power, and conflict in resource-dependent regions.

Part I: Introduction

In our second introductory chapter, we provide a foundation to guide the book by introducing the core concepts of social geography and political economy. In this chapter, we explore debates that define social geography through the reflexive and interactive relationships between social processes and relationships that shape space and place. We also outline a political economy framework that considers how important topics, such as fluctuating markets, the evolution of rights to property and resources, political relations, and capitalist production strategies that shape labour restructuring, impact resource-dependent regions.

Part II: Organizing structures

Part II consists of three chapters that portray the restructuring of local and global capitalist economic systems that concomitantly shape the challenges and opportunities for resource regions in this global capitalist economic system. To start, Chapter 3 describes the basic structure and functions of the capitalist economy. We situate resource-dependent regions within this capitalist economy, with a particular focus on how political and industrial restructuring shifted the organization of labour, modes of production, and development in these rural regions as a part of a broader global capitalist strategy. This provides an opportunity to introduce issues prevalent in resource-dependent regions, such as patterns of uneven development.

In Chapter 4, there is a more detailed approach to situate resource-dependent regions within the global capitalist system as they are shaped by international trade agreements, international institutions, and related international processes. The topic of 'scale' is used to situate multi-national industries in a global system where firms reorganize their access to materials, labour, production, and distribution in order to improve their competitive position. Drawing upon Wallerstein's world systems theory, and supplemented with other concepts such as 'core–periphery' and 'heartland–hinterland', we are able to introduce the dependent relationship between resource regions and the global capitalist economic system.

The last chapter in this section builds upon the notions of dependency introduced in Chapter 4. In this case, however, we focus on the unique context of the resource-producing regions themselves. Staples theory provides a useful approach to understand the dependent relationship between export-oriented resource-dependent regions and the vulnerabilities that arise from their participation in the global economy. Other theoretical frameworks, including evolutionary economic geography, institutionalism, and environmental economic geography, then provide a more contemporary approach to situate resource-dependent regions in renewed multi-national firm strategies that affect the resilience of these regions.

Part III: Mobilization through institutions

Part III consists of six chapters that portray how different institutions and institutional actors are shaping expectations and processes of change in resource-dependent regions. Chapter 6 begins by introducing the concept of institutionalism and situates its actors and structures within post-war Keynesian and Neoliberal political economy frameworks. We also include attention to techno-economic paradigms that have functioned through changes in the global economy. Despite the transition to Neoliberal politics, state regulations and interventions have been called upon in many international examples to correct market imperfections and capital failures resulting in a blending of Keynesian and Neoliberal policy approaches. The remaining five chapters then focus on the strategic actions of these actors, including industry and business, senior governments, local governments, local service providers and voluntary groups, and individuals.

Chapter 7 introduces the restructuring of industrial strategies over time with an interest in the scope and scale of company structures and investments in order to improve their global competitive position. We discuss three strategic trends deployed by big businesses through their pursuits of cost minimization, risk sharing through sub-contracting, and global reinvestment to more strategic and low-cost producing regions. The chapter then turns its attention to the restructuring of big labour that was once an important source of plentiful high-paying jobs and political power in resource-dependent rural and small town places. As industry restructured their operations, capital has moved to low-cost regions that have fewer regulations, lower levels of taxation, lower levels of unionization, and cheaper labour; resulting in significant job losses and weakened unions in developed countries. Similar to their corporate counterparts, however, we also trace how unions are also merging and scaling up in order to remain viable. With industry changes to procurement strategies and reduced employment, we finish the chapter by discussing the implications on the small business sector in rural and small town regions.

The restructuring and contradictions of senior government Neoliberal policy approaches becomes a central focus in Chapter 8. Despite strategic post-war investments that were instrumental to developing hinterland resource regions, we detail the restructuring and closure of many senior government programs and services that weakened much needed supports to generate renewal, innovation, and transformation in resource-dependent regions. As senior governments sought to reduce government expenditures, they disengaged from small communities and undermined their own capacity to detect the early warning signs of challenges confronting resource regions within this Neoliberal political economic landscape. With little change to the focus on resource exploitation, we describe how senior government policies became trapped and weakened in their pursuits to sustain existing and alternative forms of resource revenues to resolve short-term cash flow pressures in such 'addicted' economies (Freudenburg, 1992).

In Chapter 9, our attention shifts to local government challenges and their responses to the accelerating changes from both industry strategies and Neoliberal policy approaches as they shape or constrain the tools available to support community renewal. Topics of capacity, community readiness, new regionalism, and new governance arrangements situate some of the critical governance issues in resource-dependent regions in a context where the repositioning of local government is challenged by outdated policies, structures, and legislative arrangements. We also discuss the maturity of local government strategies in Neoliberal political economies as they move to strengthen the competitive advantage of place-based assets instead of chasing funding programs that do not always strategically align to support the transformation of resource-dependent communities.

Chapter 10 concerns changes to civil society groups as they have been impacted by Neoliberal policy shifts and more flexible productions regimes in resource-dependent regions. As such, the first goal of this chapter is to situate services and voluntary groups in resource-based places within broader Neoliberal restructuring strategies that have significantly reduced supports and offloaded public services to contractors in the private and civil society sectors. As senior governments work to reduce expenditures and communities increasingly experience the fallout from industry strategies that have shifted production to low-cost regions, we also detail the increasingly important role and leadership that voluntary organizations are exhibiting to support community development and economic development pursuits in what can often be volatile contexts. This has not been an easy task, given the unique challenges that a mature staples-dependent environment can impose on the structure and operations of such organizations. With the loss of both industry and senior government supports, these types of organizations can feel pressured to deliver services beyond their human, fiscal, and infrastructure capacity. Drawing upon a community capital framework, the chapter wraps up by detailing the practical responses that civil society groups are putting in place in order to improve the resilience of their operations as they seek to diversify both their human and financial resources, as well as their use of strategic partnerships and smart infrastructure to support renewal.

The final chapter in this section is about the different ways that the capitalist political economy of resource-dependent places has impacted individuals through the restructuring of work, household experiences, and aging. In this respect, our focus is on the changing nature of work, commuting, identity, and social and support structures. Four different types of individuals are profiled throughout our discussion, including the worker, children, women, and older residents through their lived experiences and well-being in these places.

Part IV: Change, power, and conflict in resource-dependent regions

Our last section turns to the future of resource-dependent places as they continue to be shaped by change, power, and conflict in the global capitalist political economy. These forces combine with important implications for equipping places with the capacity to successfully pursue transformations from an old economy to a new one driven by entrepreneurialism, innovation, and competitive place-based assets.

In Chapter 12, we portray how change and uncertainty play out through power and conflict constructs at the local level. As fear can be a product of change and conflict, we discuss how it affects the stability of local labour, businesses, and government stakeholders as they work hard to preserve the status quo and viability of their current position and assets in the community and global economy often at the expense of pursuing innovation and transformation. The latter half of the chapter then turns to Neoliberal public policies that have undermined the political capital and renewal of mature staples-dependent regions as they have also focused on preserving short-term revenues and responding to corporate interests in this global capitalist political landscape.

We continue our theme of change, power, and conflict in resource-dependent regions in Chapter 13. In this case, our focus is on the political economy of resource town transition through processes of continuity, transition, and emergence as they play out through economic, demographic, social, and policy changes by local, state, and industry stakeholders. Fluid capital, mobile labour, Indigenous rights, and environmentalism all become important components in this complex landscape. The strategic actions by these stakeholders and rightsholders is instrumental in shaping whether or not resource-dependent rural and small town places remain stuck in the old economy, or whether they are able to re-imagine and strategically support place-based assets so as to break the path dependent trajectories that are so common with staples-dependent economies. Neoliberal policies, however, are reducing their capacity and leverage to re-position resource-dependent regions in the global economy in response to rapidly changing market conditions.

Before offering concluding remarks on our contribution, as social geographers, to the political economy framework of resource-dependent regions,

Chapter 14 represents a reflection of socially constructed spaces in these contested landscapes. Indigenous rights, gender, and aging provide focal points for our inquiry. As resource-dependent regions pursue economic diversification to break path-dependent cycles, we also reflect on the potential of amenity migration, creative class, and services to support age-friendly communities for the next generation workforce and retirees in these renegotiated landscapes.

Closing

While rural and small town places have always been immersed in change, the pace of change under processes of economic and political restructuring has accelerated. Dynamic change and transition is now underway across the social, economic, political, cultural, and demographic fabric of remote, rural, and small town communities in developed economies. Our attention will be on understanding how communities can develop based upon the circumstances and assets of their 'place', as well as how they may need to scale up to work as regions in order to realize their aspirations. The reason for the dramatic acceleration of the pace of change, and the reasons for the re-orientation of community development to places and regions, is rooted in the trajectories of both the economy and its political superstructures. It is to a description of these trajectories that the next part of the book is dedicated.

2   Social geography and political economy

Introduction

As noted in the introductory chapter, this book explores the political economy of resource-dependent regions as it developed during the decades after the Second World War and how it is shifting still at the start of the 21st century. The scope is limited to such regions in developed economies and builds from a foundation of our work in the Canadian context. Our approach is rooted in the concept of political economy as a framework for understanding both context and change. A social geography perspective for understanding both political economy and the people and communities that make up resource-dependent regions also frames our approach.

This chapter sets out a foundational discussion of the core concepts of social geography and political economy. This is important because it frames not only how we will explore the topics, but also delimits the topics that we will and will not cover in that exploration. While a wide range of researchers, from a host of disciplines, employ a general political economy framework in their studies, our social geography approach casts a specific orientation to our work. Before moving into the second part of the book, it is important to set out a third framework necessary for understanding the social and economic development of resource-dependent regions. That framework concerns colonialism.

Geography: A perspective

The discipline of geography has roots that link back to antiquity. Since that time, it has continued to play a central role in the constellation of disciplines found in our contemporary universities. As the world enters the 21st century, and a period of accelerated and dynamic local, regional, national, and indeed global change, geography has much to offer as a perspective for organizing our understanding of those contexts and that change.

Like other social science disciplines, geography has changed over the decades in response to social, economic, and political climate of different eras. Like those other social sciences, it has been something of a mirror reflecting the transformations of societies through the years. As a result, students interested in the history and development of the discipline have a rich literature to which they can turn for more insight chronicling the debates and transformations of geography and geographic inquiry (Gregory, 1978; Livingstone, 1992).

For our purposes in this book, we start by recognizing first and foremost that the discipline of geography is as much a perspective as it is a body of research and theory. It is a way of looking at the world, a way of making sense out of the complexity of the world. This perspective creates an opportunity for understanding, and for interpreting, the changes that we see around us.

Traditionally, geography has examined a reflexive or interactive relationship – namely the role of the natural environment in limiting and creating opportunities for people, and the role of human societies in shaping and changing their habitat. While emphasis in the study of this reflexive relationship has shifted over time, the connection between communities and places deeply informs this book.

As a result, we can understand human geography as involving a number of facets in the changing relationship between societies and places. First, there is the study of 'patterns in places', or of distributions. This is an older geographic approach that can still contribute to building understanding. There are also opportunities to study the role of cultural, economic, political, and social forces at work in shaping the landscape (natural, constructed, and metaphorical). We take up an overt focus upon 'place-based' community development as the touchstone for our work on the transitions facing many resource-dependent communities in remote, rural, and small town regions.

As we build on the notion of geography as a perspective for understanding the world around us, and specifically for understanding the changing relationships that shape communities and places in resource-dependent regions, we need to add two other observations. The first has to do with the notion of time. With an interest in the place-based study of community change, it is critical to set all our inquiries into the context of time, of history. There are many reasons for this. To start, land use and community development patterns are created and re-developed over time. Since there is a very real momentum to decisions and actions taken in the past, they will have lasting impacts as the consequences continue to play out in the future of these places. Therefore, we cannot forget that our studies of places, and the processes shaping these places, will be both time- and location-specific.

The second observation has to do with the notion of geography as an integrative discipline. When we look at the social science disciplines, most are what we could call 'systematic' sciences defined (for the most part) by the object they study. But the social science study of geography is somewhat different, it is more of an 'integrative' science. With a focus on places or regions, the goal is to see how separate forces, issues, or objects of study interact to shape those places or regions. History uses time as its metric for understanding changes, issues, forces, and outcomes. Geography in turn uses spatial setting and spatial relations as its metric. As a result, through the normal course of its work, human geography will draw upon the insights of a host of disciplines (objects, studies, theories, explanatory frameworks, etc.) and it will contribute its own spatially informed perspective to enhance understanding of the synergistic or cumulative processes and outcomes. In other words, human geography is often about testing theory in places.

The value underscoring the testing of theory in places was well developed by Doreen Massey in her book *Geography matters!* She wrote about how general processes "never work themselves out in pure form. There are always specific circumstances, a particular history, a particular place or location. What is at issue is the articulation of the general and the local" (Massey, 1984a, p. 9). In the case of the resource-dependent regions, Argent (2017b, p. 18) argues that "due to a mixture of differing histories, initial resource endowments and institutional frameworks, the global economic landscape strongly resembles a mosaic, a patchwork of strongly and subtly contrasting hues, even where capitalism is the dominant paradigm for organizing an economic system. History and place matter!" We agree, and this motivates the book's attention to resource-dependent places.

Social geography

If human geography can offer a perspective on the world, we employ a social geography framework in our exploration of the political economy of resource-dependent places and regions. Social geography is a relatively recent addition to the suite of sub-disciplines within human geography, and its emergence is tied to both the historical trajectory of human geographic thought and to the specific social transformations of the 1960s and 1970s in developed economies.

Building upon long-standing interests in human geography, social geography focuses on the reflexive or interactive relationship between society and space. There are a number of very good introductions to social geography that provide a more detailed portrait of the emergence of this sub-discipline and which situate social geography within the dynamic transformations of both social science and human geography. That said, definitions of social geography are a challenge. To say the least, there is a lack of agreement about an overall definition. At its simplest, however, it could be said to be interested in the analysis of social phenomena and processes across space and in place. Cater and Jones (1989) argued that social geographers were applying a spatial perspective to an otherwise interdisciplinary pursuit of understanding the social production and re-production of places and spaces. This created strength in allowing the flexibility to draw from other disciplines as needed, so as to explore the meanings and implications of social relations, actions, processes, and outcomes.

Cater and Jones (1989) were, however, concerned with whether social geography would be seen as a relevant sub-discipline outside of academia. As such, they argued that "the vital strand of social geography is its potential relevance to contemporary questions of public concern and political debate"

(Cater and Jones, 1989, p. ix). The study of such issues was not to be just about the facts (as older geographic traditions had done) but on theoretical explanations and a wider social concern for who gets what, where, how, and why. The 'what' aspects of the inquiry focused upon issues such as work, housing, crime, gender, racial and ethnic minorities, urban neighbourhoods, and rural society. To explore the 'why' aspect, Cater and Jones argued for increasing attention to theory and critical theoretical debate. In their exploration of these issues, they drew upon a number of political economy theories.

In her 2001 textbook, Valentine continued with the observation that social geography emerged as a distinct sub-discipline during the socially and politically turbulent 1960s and 1970s. Engaging with issues of social inequality and social justice, Valentine highlights the engagement with Marxist and feminist theoretical frameworks – which were concomitantly emerging within the social sciences more generally over that period.

Updating the disciplinary history and legacy from Cater and Jones, Valentine (2001, p. 1) identifies how "in the late 1980s and 1990s this sub-discipline was influenced by the 'cultural turn' in geography, leading to a shift in emphasis away from issues of structural inequality towards one of identity, meanings, representation, and so on". With an explicit goal to remain close to the interdisciplinary roots of social geography, Valentine argues that her text:

> makes no claim to occupy a discrete intellectual space which can be identified or sealed off from other traditional subdisciplinary areas such as cultural geography or political geography. Rather, the plural social geographies which emerge here are a porous product – an expression of the many connections and interrelations that exist between different fields of geographical inquiry. Indeed they are perhaps more appropriately characterized by the subtitle: *space and society.*
>
> (Valentine, 2001, p. 1; italics in original).

With this interest, Valentine takes a different approach to the issues-based approaches of earlier social geography texts. Instead, the focus is on the 'sites' where social processes and relations work themselves out: the body, the home, community, institutions, the street, the city, the rural, and the nation. Through these sites, and against the backdrop of ongoing theoretical debate, understandings of the changing social construction of places, spaces, identities, processes, relations, and outcomes can be explored, compared, contrasted, and debated. Fixed representations or explanations are not adequate, because that lived world is continuously in a state of change and flux. For resource-dependent rural and small town places, these circumstances of continuing change nominate it as an ideal site for a social geography-based political economy study.

From its inception then, social geography has been interested in social theory. As the social sciences engaged in a ferment of theoretical debate, social geography participated and was transformed. Our choice of a social geography framework for this exploration of the political economy of resource-dependent places and regions draws inspiration from these threads and is founded upon a number of key strengths. First, is recognition that social space is holistic. While we must study the world in segments to better understand its complexity, our challenge is to reassemble the diverse and diffuse research on those individual segments in order to get a better portrait of the 'whole' of these social spaces. Second, social geography supports and countenances a wide plurality of approaches, and of a tolerance for this plurality, for studying social geographic issues. We have found these strengths to be both useful and important.

Political economy

While theoretical debate can seem intimidating to some, we take a rather more simple approach – that a theory is nothing more than an idea as to why things occur or why they are; an idea about why aspects of the world work as they do. From that simple understanding, we gain an orientation to the things each theoretical framework thinks is important – the issues they focus upon to guide and inform their explanations.

In this section, we explore the general notion of theory and introduce the framework of political economy. We close the section by also introducing some aspects of a Marxist political economy, as its critique of the capitalist economic system provides topic areas that we can explore around the changing nature of work and economies in resource-dependent regions.

Theory

Beginning with the notion that a theory is just an idea – an idea that tries to describe how aspects of the world work – we need to recognize several additional points. First, there may at any time be numbers of competing ideas, or theories, about how aspects of the world that we are interested in work. Different ideas, or theories, may point to different indicators or issues as being the ones that are more important in organizing or explaining how things work. These are the foundations for debate and dialogue. At any given time, different theories may posit different causes and explanations for the changes that we see around us. They may also posit different pathways or focal points for solutions. Over time, we can hopefully learn from the insights that these different theories have to share – look for points of commonality and points of difference – try to see how they collectively help better describe the world around us.

A second point is that, over time, theories themselves shift and change. Writers confront new information and evidence. They draw in new ideas or let old ones drop away as time and society change around them. Thus, the debate and discussion about theories of change and transition are themselves fluid. The problem for understanding comes when advocates become too entrenched in particular ideas and theories that they not only close themselves off from the insights of other theoretical frameworks, but they close themselves off from innovation and new insights within their own theoretical framework. It is our hope that the contribution of this book is one that is open to insights from a host of theoretical frameworks and contributes, in turn, to innovation and insight in the development of better understandings of resource-dependent regions.

A third point is that theory organizes all aspects of our work. Whether it is research, politics, or practice, we are all informed by our own ideas and understanding of how we think things work. As a result, it is important that we recognize overtly our theoretical framework so as to be clear about why we are making certain links and connections in observing and interacting with the world around us.

A final point we wish to raise, and one that challenges many with respect to the word 'theory', is the notion of scale. At its simplest, theory can be thought of as working at a minimum of two scales. At the highest, or 'grandest' level, theory can talk about the organization of society. Some examples of theoretical frameworks at this level include political economy, Marxism, feminism, and neoclassical economics. Each posits a series of organizing features that they suggest structure the organization and functioning of entire societies. While the theories themselves may not be especially 'observable on the ground', adherents point to outcomes as evidence of the workings of their theoretical framework. At this macro level, there is often complex and specialized terminology and it is this complex and specialized terminology that can make these theoretical frameworks seem scary and impenetrable.

At a much more operational level, there are also theories or ideas about the relationships between observed events and outcomes. A simple example might be housing. This could include the relationship between income and housing affordability, or the relationship between interest rates, new home ownership, and the ability of different households to qualify for financial assistance in home purchasing. Like with grand theorizing, these ideas or theories of how things work can have specific terminology and can direct us to look only at certain issues and to structure our research and our understanding.

Political economy

A wide range of social scientists consider themselves to be working within a political economy framework. For some, this framework is refined by looking specifically at, for example, a Marxist version of political economy. In this book, however, we remain at the broad and general level.

In very simple terms, a political economy framework articulates an understanding of society as a complex and interactive entity. It does not legitimize purely political, purely social, or purely economic explanations. Instead, it recognizes that these elements interact and are, therefore, not divisible. By way of illustration, Sheppard (2015, p. 1113) recently, argued that capitalism "cannot be understood, or practiced, simply as an economic process; its economic aspects are co-implicated with political, cultural (gendered, raced, etc.), social, and biophysical processes, in ways that repeatedly exceed and undermine any 'laws of economics'".

This 'indivisibility' creates challenges. One of the first is that a holistic approach to the understanding and study of all issues at all times is almost impossible, given the complexity involved. Second, when we divide out the component parts of a political economy framework for more detailed individual study we must remember to reassemble them again with our new understandings. Nozick (1999, p. 5), writing about issues of community, social, and environmental sustainability argues the need for "an *integrated* approach that addresses economic, ecological, political, and cultural development" (italics in original).

For our purposes, a political economy framework is a useful and broad umbrella beneath which we can shelter our explanations of the changing economies, communities, political structures, cultures, and demographics of resource-dependent places and regions. Each of these topic areas has practical level theory that helps inform why elements are changing and transforming. But for us, it is how the interactions are creating a particular type of change and transformation that is generally unique to resource-dependent places and regions that underscores the value of a political economy framework.

Another important aspect of a political economy framework is recognition that all explanations are conditional or contingent. That is, the constituent elements are subject to change over time. Transformation is dynamic and thus we would not expect the understanding of resource-dependent regions created in the 1950s to bear much resemblance to the understanding of these same places at the start of the 21st century. A second element of the conditional nature of explanations is that they may also vary across places. The history of places together with their contemporary socio-political-economic structure creates key differences in how more generalized pressures and processes of change unfold. Just as we would not expect resource-dependent places from two different eras to look and act the same, neither would we expect two resource-dependent places from very different regions to look and act exactly the same.

In his fine essay on political economy in the *Dictionary of human geography*, Sheppard (2009, p. 547) describes it first as the "study of the relationship between economic and political processes". Tracing a lineage of interest in the confluence of these processes from Aristotle, through French and English theoreticians, Sheppard highlights the emergence of constituent topics of interest and study within political economy. These topics include industrial development, the division of labour, the role and influence of markets, the role of property and property rights, and the many facets over time of trade and trading relations in structuring economies, political relations, societies, and communities.

Sheppard also describes the intellectual 'split' between Marxist political economy and liberal political economy schools. A Marxist political economy approach focuses upon how "the crucial relationship between political and economic processes is the way in which capitalism engenders, and is shaped by, political struggle between classes" (Sheppard, 2009, p. 548). In turn, a liberal political economy approach adopts a more neoclassical economic mantra to how capital and labour act out their relationships according to the processes of free markets. However, the liberal political economy approach is complicated by the debates between more progressive and more conservative advocates. The more progressive approach came to be associated with the work and writings of Keynes and the principal "of continual state intervention to manage the contradictions of capitalism to the benefits of the nation and its least well-off citizens" (Sheppard, 2009, p. 548). The more conservative approach has been transformed in recent years to support the broader Neo-liberal project, which "pays close attention to political processes, but only because such processes tendentially undermine what it regards as the proper functioning of the capitalist economy" (Sheppard, 2009, p. 548). Both Keynesianism and Neoliberalism are discussed later in this book.

For DeFilippis and Saegert (2012a), political economy is conceptualized within the capitalist economic system. With our focus on resource-dependent places in developed economies, this focus is again relevant. For them, communities and individuals function within the opportunities and constraints of a capitalist political economy. By this, they mean the tangle of social and economic relations that affect the local / personal, but "exist well beyond" those constrained geographies (DeFilippis and Saegert, 2012a, p. 4). For Sampson (2012, p. 315), these wider limitations support the contestations and conflicts of change at the local level. This invites the question: to what degree are community problems really community problems or are they problems of the capitalist political economy that communities must confront? In other words, "community development emerges in the context of the current limitations of the capitalist political economy to fulfill the needs and desires of the community" (DeFilippis and Saegert, 2012a, p. 5). Resource-dependent places know this question well, as ongoing crises around corporate profitability have driven dramatic change in local industries and the communities connected to them. To help situate a general political economy framework for understanding remote, rural, and small town community development within the wider structure of a capitalist economy means introducing a discussion of Marxist political economy.

Marxist political economy

As noted above, one strong thread of political economy thought is the Marxist critique. In keeping with our approach to theory as simply a body of ideas about how the world around us works, a Marxist political economy framework provides a critique by which to understand the capitalist economy and our capitalist society. After the transformation from mercantilism to capitalism as the organizing basis for economic exchange, Marx undertook to examine and elucidate the workings of this new economic system. His main objective was to characterize and understand the internal or underlying structure of the capitalist economy (and thus capitalist society). The aim was to discover the causal mechanisms central to the underlying structure. Given that resource-dependent remote, rural, and small town places in developed economies are part of the global capitalist economy, a first step to understanding past trends as well as current and future pressures, is with a carefully developed characterization of how that capitalist economy works.

As fits its roots as a holistic political economy framework, Marxist theory attempts to understand society as a whole, in that it is more a general theory of society than an approach to simply understanding the economics of exchange. Marx argued for the need to understand two key characteristics of the capitalist economic system. The first is structure. By this he meant the class-based division of labour and society – with this class division working to organize or structure all manner of social, political, and economic relationships. The second is the causal mechanism driving the organization of relations within that structure. In this case, it is the capitalist mode of production – a mode of production that sees capital combine resources with labour to create a product that can be sold at a profit to capital. The profit imperative is key and will be discussed again later.

Describing the causal mechanisms underlying the capitalist economy, and capitalist society more generally, raises the theoretical delineation of the 'concrete' and the 'abstract'. In trying to characterize and understand the internal structure of capitalist society, Marx identified the need to differentiate the apparent and observable features of social / economic life from the underlying structures. To make sense of the system, there was a need to first identify these underlying structures, or causal mechanisms, that are the foundations for the structures of everyday life and economy. A Marxist social analysis of the capitalist mode of production, therefore, consists of at least two stages. The first is the 'abstract' – a study of the internal structure of the capitalist mode of production. This level of analysis is theoretical. The second is 'concrete' – which involves a theoretically informed empirical analysis of actual social and economic forms.

Brought into geography in the early 1970s by writers such as Harvey (1975), a Marxist political economy approach sought to bring together the social and the spatial. With an emphasis upon production in the capitalist economic system, the focus turned towards the 'production of space' (Levebvre, 1991). More specifically, it focused on the uneven or inequitable production of different spaces. These produced spaces are the places where our everyday lives are acted out. But they are not just a stage upon which the acts of a capitalist economic system play out, indeed these places do themselves play a role in affecting how the social and spatial configure and re-configure over time and through the successive crises of the capitalist economy. In the next part of the book, we explore in more detail concepts associated with the nature of capitalism, and the global structure of capital, and their impacts on resource-dependent regions.

Colonialism

Exploring the social construction of places, economies, and regions through a social geography lens creates an opportunity for us to include the impacts and implications of historical processes. A political economy approach also demands that the interplay between historically constructed, and currently contested, forces (economic, political, social, cultural, etc.) be considered in their specific place- and time-bounded contexts. Both approaches consider this important, because the legacies of different historical processes will continue to impact both the current situation and future possibilities of these regions. While different historical processes may have more or less significant, or more or less lasting, impacts and implications, one of the most significant and lasting of these is colonialism.

Colonialism is an attribute of many resource-dependent rural and small town places and regions within developed economies. As will be described in the upcoming chapters, the post-Second World War expansion of the industrial resource-development model into rural and remote regions was often designed and supported by the planning and policy actions of centralized state governments. In numbers of jurisdictions, the understanding of rural and remote regions by urban-based policy makers was that of a largely empty, unsettled, resource rich 'bank' which could be exploited to help develop local, regional, and national economies (Tonts et al., 2013). However, many of these targeted development regions were not empty or unsettled at all. Many were home to various Aboriginal / Indigenous populations. These Aboriginal / Indigenous populations had often lived in these regions for centuries, if not millennia. They continued to live in these regions throughout the expansion of the post-war industrial resource model – and they continue to live in these regions still.

While not the case in all jurisdictions, there are many examples where industrial resource-development policies moved forward without care and attention to, or consideration of, Aboriginal / Indigenous populations, their cultures, and their economies. Again depending upon the case, this often included proceeding without attention to the rights and title to the land and its wealth that Aboriginal / Indigenous populations held. McDonald (2016, p. 93) describes the case within Tsimshian territory in what is now northwest British Columbia:

> Beginning with the distorting codification of Tsimshian property concepts during the establishment of the Indian reserve system, property relation-ships were even more radically altered in the twentieth century. The Canadian government moved from assuming legal ownership of the Tsimshian territories to implementing economic control by developing a set of provincial and Dominion laws foreign to Tsimshian values. These laws increasingly brought the resources of the regions under the explicit control of Dominion and provincial governments and the interests they

represented. This made it difficult for the Tsimshian to maintain their systems of resource management and exploitation. In addition, the feast was banned by the Indian Act, interfering with the Tsimshian legal procedures for maintaining internal control over landed property and resources. This transformation from Tsimshian lands to Crown lands was accomplished in little more than four decades after the establishment of the first HBC post – a single lifetime.

In many Indigenous areas, this loss of control of lands and resources has had profound social, cultural, and economic impacts. It has often made it difficult for people to stay in the local communities where they were born. One of the reasons for this is that people are getting less access to make a living off traditional resources. Pedersen (2015) describes the processes of state-led exclusion impacting Aboriginal / Indigenous peoples. Using the example of the Sámi in the north of what is today Norway, Sweden, and Finland, he argues:

> What in fact has happened within the management of fisheries is that a large part of the coastal Sámi population, along with other small scale fishermen in the same region, gradually have lost the right to make their living off the traditional local and regional fish resources. They have always pursued a sustainable fishery, but this ideology of sustainability gave no reward when a new fishery regime was introduced. In 1989, those who had acted according to the principle of not overexploiting the cod stock were excluded from getting their fair share when the so-called vessel quotas were introduced. They had not fished enough!
>
> Since then the quotas have been made tradable, privatized, and thereby increasingly become a private profit making commodity. According to the Sámi Parliament of Norway, established in 1989, those regulations have disregarded both customary and Indigenous Sámi rights to traditional fisheries in the local waters.
>
> (Pedersen, 2015, p. 193)

These post-war processes of displacement and 'reterritorialization', of course, occurred in contexts layered with much earlier, and often multiple, processes of displacement and reterritorialization (Harris, 2002). The processes of displacement and reterritorialization wrought by colonialism had devastating consequences. From economic marginalization, to separation from traditional territory and traditional land use activities, to purposeful undermining of traditional governance and cultural practices, all these created a lasting imprint. Drawing upon work in Australia, Sheppard (2013, p. 272) described how its legal structure was developed to enable

> land-based resources to be alienated from Indigenous inhabitants, reflecting the racialized attitudes accompanying European settlement constructing these as blank spaces awaiting alienation for proper usage

under the aegis of market mechanisms, but also more recent shifts recognising Indigenous land rights, and compensatory payments and employment schemes in the mining industry.

In Canada, residential school practices aimed at assimilating Aboriginal people into the white settler society removed Aboriginal children from their communities and traditional territories, and forbade those children from speaking their traditional languages or participating in their traditional ways of life. The Royal Commission on Aboriginal Peoples (1996) identified terrible patterns of abuse whose individual and collective harm still requires healing. In 2008, a national Truth and Reconciliation Commission was appointed in Canada to examine the residential school issue. The preface of their final report was blunt:

> Canada's residential school system for Aboriginal children was an education system in name only for much of its existence. These residential schools were created for the purpose of separating Aboriginal children from their families, in order to minimize and weaken family ties and cultural linkages, and to indoctrinate children into a new culture – the culture of the legally dominant Euro-Christian Canadian society … The schools were in existence for well over 100 years, and many successive generations of children from the same communities and families endured the experience of them. That experience was hidden for most of Canada's history, until Survivors of the system were finally able to find the strength, courage, and support to bring their experiences to light in several thousand court cases that ultimately led to the largest class-action lawsuit in Canada's history.
> (Truth and Reconciliation Commission of Canada, 2015, p. v)

Addressing the longer-term consequences of colonialism has only just been initiated in many jurisdictions. Recognition and apologies are a start, but attention to long-term capacity building, support for self-governing, and attention to (re-)building cultural, spiritual, and personal and community health and wellness are all necessary.

To the longer-term consequences, there are also the direct impacts of resource-development activity that need to be considered. These include the social and cultural effects of large numbers of workers coming into the region and the issues of alcohol or drug abuse that can accompany 'boom' development, pressures on the land base, racism and discrimination, etc. As noted by O'Faircheallaigh (2013, p. 20):

> Indigenous peoples have been especially susceptible to marginalization and destruction of livelihoods, because they rely heavily on land and resources that are susceptible to environmental damage from resource extraction; are vulnerable to the impact of immigrant populations; and

lack political influence because of their small numbers combined with discrimination and social disadvantage. Indigenous people frequently live in poverty adjacent to ... enormous wealth.

In many resource-dependent rural and small town regions, however, there is increasing recognition of Aboriginal / Indigenous rights and title. In some jurisdictions, this comes as a result of court cases. Regardless, there is now increasing attention to inclusion of Aboriginal / Indigenous interests. In Australia, legislation governing the recognition of access, ownership, and compensation for resource-development activities is in place in numbers of jurisdictions (O'Faircheallaigh, 2013). In Canada, rights of consultation and accommodation have been defined by courts and are now generally part of the legal approval and project review processes (Gillingham et al., 2016). This has meant that community impact benefit agreements are increasingly being negotiated with Indigenous communities, especially in regions with unresolved land claims (Storey, 2010). These arrangements are providing Indigenous communities with more contract and joint venture opportunities in emerging resource-development projects, largely through camp and catering contracts, but also through security, transportation, maintenance, and other logistics. As many such processes move forward, there is agreement that some resource-development proposals can create the potential for Aboriginal / Indigenous peoples to "maintain their existing livelihoods or choose to supplement these with different forms of economic activity" (O'Faircheallaigh, 2013, p. 28). However, the confidential nature of these agreements means that consideration of other Indigenous and non-Indigenous interests may be excluded, thus putting the transparency of equal distribution of benefits into question (Fidler and Hitch, 2007). Further, there are also important questions about the roles of such legal and legislative changes, and of court-recognized rights and title, in those cases where Aboriginal / Indigenous communities do not support particular forms of resource development, and where they do not wish those developments to proceed.

Closing

The purpose of this book is to explore the political economy of resource-dependent places and regions, within developed economies, at the start of the 21st century. This chapter has framed the perspectives that we will bring to this purpose. It is shaped by our interest in the geography of recursive relationships between people and the places where they live and work. It is especially informed by a social geography framework that focuses upon the socially constructed forces that shape both societies and places over time. To this, we have added a more general political economy concern for the indivisibility of social, economic, and political processes and explanations. Our commitment, therefore, is to a more holistic approach to understanding the trajectories of change impacting these resource-dependent regions.

By describing the political economy of resource-dependent regions, we wish to cast light on the elements and the processes driving change and transition. We want to especially highlight those forces internal and external to such regions, and to also highlight the more general aspects of these changes and transitions that may sometimes be obscured by the idiosyncrasies of unique places or regions. Across OECD countries, resource-dependent regions do find themselves enmeshed in common challenges. To build awareness and understanding can be a first step in assisting researchers, policy makers, elected officials, local volunteers, residents, workers, business owners, and communities to better engage with the opportunities and challenges of change and transition.

But our interests are not just with building understanding. Our interests are also with engaging proactively with remote, rural, and small town resource-dependent places and regions. Moving research into action, and supporting people and places through change, has been the motivating part of our work for nearly two decades. We agree with Del Casino (2009, p. 27), who argued for a more critical political economy, where 'critical' means "being engaged in a project of understanding inequality and difference, however, in the hope of producing knowledge that will, in some small way, create positive change". It is with this foundation that we can move into the next part of the book, which describes some of the fundamental organizing structures of resource-dependent regions.

Part II
Organizing structures

Introduction

As described in the first part of the book, our purpose is with developing a critical political economy framework for understanding the trajectories of change that are impacting remote, rural, and small town resource-dependent regions in developed economies. Building from a social geography perspective, and an activist interest in place-based community development, we recognize that there are important 'structures' in place within which the actions, opportunities, and challenges confronting such resource-dependent places and regions are organized. This part of the book is directed at describing those broader organizing structures.

Part II is divided into three chapters. Each of the chapters addresses elements of how the capitalist economy is organized at the local and global level. The purpose is to identify those characteristics that are important if we are to better understand the dynamics of change in resource-dependent places and regions.

In Chapter 3, we continue our introduction to the capitalist economy by describing its basic form and structure. This introduction, started in the previous chapter, now digs into more of the mechanical details, structures, relations, and imperatives of the system. To do so we draw especially upon a Marxist critique of capitalism and focus upon such topics as the nature of capitalism, the organization of different modes of production, and the spatial division of labour.

Given that 'change' is the organizing part of our interest, the last part of Chapter 3 turns attention to a description of the changes commonly described in resource-dependent regions and to theoretical explanations for those changes. Particular focus is put on the notion of 'restructuring', including the arguments for and against this form of change as a fundamental shift in the organization and nature of capitalism.

Chapter 4 changes the scale of inquiry. After delineating the structural and relational elements of a capitalist economy, there is a need to bring scale into the equation. Just as colonial empires and mercantilist trading alliances were global in reach, we know that the capitalist economy is globally organized.

Multi-national firms organize their inputs of materials and labour globally, they manage their production and distribution operations globally and in real time, and they pursue cost-price advantages by locating and re-locating aspects of their production and marketing chains as global circumstances change. To extend our understanding of the capitalist economy by incorporating spatial aspects into a more systems-based approach, we draw upon the work of Immanuel Wallerstein and the concept of 'world-systems analysis' – especially the notions of dependency and marginality within the global economic system. It also adds other geographic elements around the notion of marginality – in our case, with particular interest in the dependent or marginal nature of resource-producing regions within advanced capitalist economies and countries.

Having introduced the capitalist economy, and the global capitalist system, Chapter 5 turns attention to the particular circumstances of resource producing regions. The chapter includes a review of examples that highlight these linkages. The core of the chapter, however, focuses on a uniquely Canadian contribution to political economy thought – 'Staples theory'. Developed by the economist Harold Innis, Staples theory looks specifically at the case of resource-producing regions that export low-value-added raw materials as inputs into the manufacturing chains of advanced economies. For places and regions that have long been organized around the export of such raw materials, or staples, this framework elucidates the continuing issues of vulnerability and dependence that such a position and role in the global economy entails. To this background, we add more recent theoretical debates, including those associated with evolutionary economic geography, institutionalism, and environmental economic geography. Finally, the chapter closes with some consideration of how the concept of 'sustainability' fits within the political economy of resource-dependent regions.

3　The capitalist economy

Introduction

This chapter explores aspects of the capitalist economy in terms of its structures and relationships. Given that resource production in developed countries functions within a capitalist economic framework, it is important to identify these basic structures and relationships. Therefore, much of the language and analysis used in this chapter draws from a Marxist political economy critique of capitalism.

This chapter is organized into two sections. The first provides an introduction to the capitalist economy. While a number of theoretical frameworks have been used to characterize this economy, we draw upon a Marxist analysis because it is aimed specifically at creating a detailed description and understanding of the components and operation of the capitalist economy. It has also worked to differentiate the observable elements of that economy from its underlying structures. This underlying structure, including the way it can be detailed, is important when we interpret change within the economic system. That topic of change is then taken up in the second section on 'Restructuring'. We use the common framework of the transition from a Fordist mode of production to a more Flexible mode of production as our starting point. After detailing the typical aspects of each mode of production, and linking those changes in resource-dependent places and economies, we then critique the concept of restructuring by asking: what has changed? While the 'concrete' or observable characteristics appear to have changed, the 'abstract' or underlying structure of the capitalist system remains very much as described in the Marxist analysis.

Introduction to the capitalist economy

Today, we have no trouble recognizing that our economy is a 'capitalist' one. That it is organized around the production and consumption of commodities, that this production and consumption is managed through processes of market exchange, and that business owners and corporate shareholders alike focus on deriving profit and capital accumulation are not new or surprising revelations. These are everyday processes that we recognize even in the organization of our own household economies.

There are a number of theoretical frameworks that have been used to describe the nature of the capitalist economy. In his essays on the nature of capital and capitalism, Watts (2009, p. 59) argues that both

> Classical political economy – Adam Smith and David Ricardo – and its Marxian critique both accepted that capitalism is a class system, [that] labour and capital were central to its operations, and that capitalism as a system was expansive, dynamic and unstable.

Most accounts describing the nature of the capitalist economy are controversial and highly debated. They are debated both within, and between, different bodies of theory. Interpretations vary, not surprisingly, on the nature of the 'relations' driving the capitalist economy.

In his significant treatment of economic geography, Barnes (1996) highlighted aspects of the theoretical debate about capitalism in the introduction to the roles that neoclassical, Marxist, critical realist, locality studies, and flexible production studies have each brought to the subject. Each body of theory first constructs its own understanding of 'reality' and then of the causal mechanisms that energize that constructed reality. In neoclassical economics, for example, the "very techniques they employed necessarily couched the economic problem in terms of maximization" (Barnes, 1996, p. 63). The project, therefore, became "to discover the most efficient (rational) actions required by consumers and producers to maximize utility and profits, respectively" (Barnes, 1996, p. 63). As he goes on to demonstrate, human complexity and the vast differences between 'products' and 'markets' make the application of models and laws derived from physical science rather ill-suited for economic geography. Instead, an analysis inclusive of the human or social relations within these 'products' and 'markets' may provide more insight. In the next section, we draw upon such an analysis.

Building from our social geography perspective, we seek to investigate the underlying structure of the 'everyday' world we live in. As a result, it is necessary to take up a theoretical literature that has long interrogated that world. In terms of the capitalist economy, Karl Marx initiated one of the most careful and long-standing of these interrogations. Recognizing that the economic organization of the world had been transformed, Marx sought to identify both the structures and the causal mechanisms for the emergent capitalist economy. The following sections explore aspects of a Marxist political economy critique as they inform an understanding of transformation and change within resource-dependent places and regions in developed economies.

A Marxist description

The capitalist economy is, of course, organized around the word and notion of 'capital'. At its simplest, we can understand capital as a surrogate for wealth. It involves both fixed capital and circulating capital (Watts, 2009). Fixed capital can be thought of as including assets such as land, machinery, etc. These things are typically long-lasting and are difficult to convert into circulating capital. Circulating capital, on the other hand, includes liquid assets (most often in the form of money or some other readily convertible form) available for spending and investment. It is to the accumulation of both fixed and circulating capital that the capitalist economy is directed.

In their textbook on social geography, Cater and Jones (1989) provide a detailed introduction to a Marxist political economy. They introduce the inherent competition between capital and labour that is embodied within capitalist production, the way that competition creates social inequality, as well as the way that competition (together with the other imperatives of capital) also creates geographic inequality. Together, their sketch connects a Marxist political economy with the path dependent nature of the economic structure of resource-producing remote, rural, and small town regions.

To address the competition between capital and labour within a capitalist mode of production, Cater and Jones (1989, p. 18) note that "We should remind ourselves that Marxist analysis of the social relations of production starts from the more fundamental proposition that it is the ownership or non-ownership of capital which determines class membership in capitalist society". In dissecting the competition between capital and labour further, they also argue that "the division of the spoils both inside and outside of the sphere of production is as much a political process as a technical one" (Cater and Jones, 1989, p. 1). As argued by Watts (2009, p. 61), that competition or division of spoils is unequal: "Marx's account identifies a fundamental contradiction of the heart of capitalism – a contradiction between the two great classes (workers and owners of capital) that is fundamentally an exploitive relation shaped by the appropriation of surplus".

In his work on a definition of capital, Watts (2009, p. 59) adds that:

> Capital is not a thing, but is a 'social relation' that appears in the form of things (money, means of production). Capital does indeed entail making money or creating wealth but, as Marx (1967) pointed out, what matters are the relations by which some have money and others do not, how money is put to work, and how the property relations that engender such a social world are reproduced.

Taken together, it is clear that social relations, and the processes of constructing and re-constructing those social relations, are central to a Marxist critique of how the relations between capital and labour are 'reproduced' over time.

Through these socially constructed technical and political processes, the social and the spatial divisions of labour are created. It is the "fragmentation of the work process both within and between individual work places and between industries; and the consequent allocation of each worker to a narrow and specialized range of tasks" that defines the division of labour (Cater and Jones, 1989, p. 16). It is by that division of labour that social inequality is created.

Geography is brought into the discussion through recognition that there is also a 'spatial' division of labour. For Cater and Jones (1989, p. 2):

> the modern production process is not only minutely broken down into a host of narrow tasks each performed by a different section of the work-force: it is also spatially differentiated in that many of these specialized tasks are highly localized, over-represented here, under-represented there, absent elsewhere.

By extension, the new spatial division of labour describes how the "new mobility of capital is related to the rise of the giant corporation" (Cater and Jones, 1989, p. 27). As economic restructuring has put pressure on the division of labour, resulting in significant labour shedding and the adoption of 'flexible' labour contracts in resource-dependent industries (something we will pursue in later chapters), so too does it put pressure on the spatial division of labour in resource-dependent regions. Structural shifts in local and regional economies can occur as a result of the replacement of older generations of technology or production activities. Cater and Jones (1989, p. 87) highlight how "new waves of industrial development do not necessarily favour the same locations as former waves". This mobility of capital has significant implications for older industrial regions – and resource-dependent rural regions in developed economies are mainly older industrial regions.

Uneven development

With respect to the mobility and the geographic patterning of labour, a distinctive contribution of geographers has been to introduce attention to spatial patterns both within and across the division of labour and how this also impacts the processes of exploitation. In this, attention has been directed to "differing patterns of uneven development – whether locally, regionally or globally" (Gidwani, 2009, p. 446).

More generally, Harvey describes how the process of geographically uneven development is tied to how "capital moves its crisis tendencies around geographically as well as systematically" (Harvey, 2010, p. 12). Connelly and Nel (2017c, p. 115) describe how this argument helps to

> clarify the reality and self-reinforcing nature of regional differences and the degree to which globalization does not create homogenization but rather in a situation in which regional and local economic advantages and disadvantages persist and can in turn be reinforced.

Structures

Marx's objective was to describe and characterize the internal or underlying structure of the capitalist economy. The search was for the causal mechanisms that reinforce and reproduce that underlying structure. As his argument unfolded, the structure was identified as the class-based division of both the economy and society, and the causal mechanism was identified as the capitalist mode of production. In this formulation, production can be thought of as much more than just the physical act of manufacturing. Linking to the preceding discussion, it is also a political process, since it organizes access to and control over resources, capital, labour, and consumption. It also organizes and regulates both rights and obligations through legal delineation. It is also a spatial process in that it organizes places and regions depending upon their functional location within production / consumption processes.

The political and spatial organization of production lends itself quite easily to patterns of uneven development. At the base of this uneven development is recognition that there will be uneven access to resources, markets, wealth, etc. Depending upon the social and political processes in place, access may be slightly more open or it may be terribly more restricted. Regardless, access must be uneven in order to establish the basic rationale and need for 'exchange' within the economic system.

If initial access to such basic elements is uneven, then the resulting systems of production will also re-create and perpetuate these geographies of unevenness. Resource-producing communities and regions, for example, typically do not control their local natural resource base. This control or the rights to access those resources are generally held by the property owner in those cases where resource lands can be privately held, or the rights are granted to industrial interests by the state in those cases where resource lands are owned by the state. With such rights granted away, resource-dependent regions are vulnerable to market and corporate decision-making. These themes are pursued further in Chapter 5.

Production requires not only access to raw material inputs, but it also requires access to labour. As already described, in a Marxist critique the capitalist economy and society are marked by a structural division between capital and labour. As a result, the uneven development that marks production also results in a spatial division of labour. Access to work, and access to certain types of work, is dependent upon location and the organization of the productive economy within that location. Again, resource-producing regions have employment opportunities that are circumscribed by the type of production. Thus, regionally concentrated economies such as resource-producing regions also concentrate particular forms of labour. To a degree, labour becomes 'trapped' in these locations, as there are typically few other local options.

In both the organization of production and the spatial division of labour, there is a significant role for the state. To start, the state sets many of the terms and conditions with respect to how capital accesses both raw materials and labour. Public policy shifts in recent decades have been especially sensitive to the shifting global economy and the competitiveness of resource-producing regions within developed economies. While these themes are further pursued in Chapter 7, at this point it is sufficient to say that these shifts continue the political imperative to support industrial interests that have long been the foundation of the capitalist economy. The state depends upon tax revenues, and since the end of the Second World War, resource industries have generated relatively large tax returns to the state – with these taxes inclusive of corporate taxes, payroll taxes, income taxes, consumption or sales taxes, and various forms of rents from the natural resource itself. So successful was this model from the 1950s to the 1980s that states became 'addicted' to the large revenue flows they provided (Freudenburg, 1992). The various crises of profitability that have marked resource industries in developed economies since the 1980s have necessitated responses from industry, of course, but they have also required changes of public policy as well in order to prop up struggling industries (Halseth et al., 2015).

In addition to regulating access to natural resources and labour, the state also manages who will be responsible for the 'social reproduction' of the capitalist economy. If we look back at early industrial development in North America, the model was the textile industry along the east coast of the United States. In these textile towns, the company was responsible for nearly every aspect of economic and social life. The company not only provided the work, but it provided the housing, schools, the healthcare facilities, it maintained the roads and the water supply, etc. Over time, and as the capitalist economy matured, many of the responsibilities previously left to industrial interests moved to the public sphere. For many of the responsibilities listed above, responsibility shifted to the state. Using tax revenue provided not only by industrial interests, but also by the workers through their income, consumption, and property taxes, the state was able to provide these needed services. This transition set up a long-term and continuing debate between capital and the state about responsibilities. Today, in many resource-dependent places and regions, companies continue to ask for concessions around taxes and other production responsibilities in order to reduce their costs and enhance their potential for profitability.

Basic concepts

There are some important basic concepts that a Marxist critique has added to our understanding of the capitalist economy. In this section, we would like to review five: 'material production', 'instruments of production', 'means of consumption', 'social relations of production', and 'historical change'.

To start, a capitalist economy is organized around 'material production'. That is, in order to add value and create the opportunity to make profit, it is necessary to transform a basic or raw material into some form of product. Of course, there can be many sequences in this process whereby the finished product from one production process may become the 'raw material' or component input for a further production process(es). This transformation is at the heart of the production process. It also provides the foundation from which profit can be derived from the sale of these transformed commodities.

In resource-dependent regions, transformation to create value (and thereby profit) is often limited to the most basic forms of manufacturing. In the forest industry, for example, processes of transformation may include:

- the harvesting of trees and their conversion into uniform length logs for export,
- the harvesting of trees and the processing of the logs into dimension lumber,
- the harvesting of trees and the processing of the logs into higher-value products such as various forms of panel boards, or
- the processing of wood waste from logging and sawmilling to support even higher value-added products in pulp and paper.

Despite the fact that these are all transformation (or manufacturing) processes, they add very limited value to the raw resource and instead rely upon volume to generate returns to capital. Similar limitations apply to other resource sectors where the raw material is minimally processed and often only involves a 'break of bulk' processing to reduce shipping costs (i.e. rocks milled into ores, or grasses threshed to separate out the seeds or grains). Throughout resource-dependent regions, the concept of manufacturing typically does not extend to complex and high value-added processes, such as those found in automobile or electronics production.

In order to manage production, the capitalist economy also needs 'instruments of production'. In general terms, there are two basic forms of instruments of production. The first is labour. Labour is the capacity to do work and it is a capacity which workers sell into the capitalist economy for remuneration. The second involves equipment. Sometimes also called fixed capital, this manufacturing equipment involves the tools that labour will use in the transformation processes of material production.

Under various crises of profitability, capital will seek to trade off the costs of labour and equipment. In some cases, strategies will explore how investment in labour-saving equipment may be used to speed up production, lower the wage (and associated employee benefits) cost, and together lower the per unit cost of the finished product. In other cases, strategies will explore relocation to regions or jurisdictions with similar economic advantages (access to raw materials, export infrastructure, etc.) but where the wage and benefits costs are much lower and thus also potentially able to achieve a lower per unit cost of the finished product. With its low value-added transformation processes, and its reliance upon volume to generate sufficient returns to capital, both of these strategies have dramatically changed older resource-dependent remote, rural, and small town places and regions within developed economies.

In his study of the economic geography of labour value, Trevor Barnes identified the theoretical justification for the second strategy. He argued that:

> If all commodities, including labor, exchange for their value, how can surplus arise, which is in turn converted into profits? Everything, after all, seems to be exchanged for what it is worth. Marx's brilliant insight is to recognize a distinction between the value of labor and the value of labor power. On the one hand, capitalists do not hire labor; what they buy is labor power, that is, the ability to produce.
>
> (Barnes, 1996, p. 60)

Shifting production from high-wage states to low-wage states increases this difference between the value of labour and the value of labour power so as to enhance profitability. As a result, "The more general point here is that the political and the social cannot be excluded from the labor process and thus from the definition of labor values themselves" (Barnes, 1996, p. 75).

With the production of commodities, the third basic concept important within a capitalist economy involves the need for a 'means of consumption'. Supporting the ability of consumers to purchase products is essential to the profit motivation that drives the capital accumulation imperative of the economic system. For example, many people consider that Henry Ford's most significant contribution to industry was the adaptation of assembly-line production technologies to automobile manufacturing – producing more cars, faster, and cheaper. Others have argued that his most significant contribution was that he paid his workers a wage sufficient enough such that they could soon be consumers of the very products they were manufacturing. In other words, he closed the loop. For Harvey (1990, p. 135) it is broader still and is linked to the political economy that emerged to support the new industrial complex phase of capitalism:

> Postwar Fordism has to be seen, therefore, less as a mere system of mass production and more as a total way of life. Mass production meant standardization of product as well as mass consumption; and that meant a whole new aesthetic and a commodification of culture.

In natural resource economies, the consumer base is not local or employed in the industry – the volume of production needed to generate sufficient returns to capital is simply too great to target such a limited market. Instead, the production is destined as inputs into other manufacturing processes and those processes are targeting large consumer markets – with these consumer markets in developed or emerging economies. Canada's forest industry developed as a supplier to the burgeoning post-war US housing market. The sheep wool industry in Australia and New Zealand developed within Britain's preferential trade system. Norway's oil and gas industry developed to provide the raw material inputs for the energy and petrochemical industry needs of Europe and Asia.

Taken together, these first three basic concepts used to describe and understand the capitalist economy create problems for resource-dependent regions and set the foundations for the dramatic restructuring experienced in the late 20th and early 21st centuries. Given the typically limited value that is added to the raw resource in remote, rural, and small town places and regions where the resource is found, and the associated reliance on volume to generate returns to capital, crises of profitability in resource industries have been met with the aggressive substitution of technology for labour or the relocation to low wage regions of the globe. In both cases, the loss of jobs, taxes, and related investments have shrunk the local benefit from local resources. The implications of an increasingly mobile and globalized capital for small and resource-dependent communities are significant. Nozick (1999, pp. 3–4) laments how

> Once-vibrant farm towns are dying as rural populations dwindle and the local farm economy is replaced by large-scale agribusiness. Single-resource towns have become ghost towns as industries shut down or move to other countries for bigger tax breaks … These are not random occurrences but the results of complex global forces working to dismantle the structures of community life … Rootlessness and dispossession are the inevitable fallout of an economic system based on free mobility of capital and global competition.

The dependence of natural resource industries on the needs of advanced manufacturing centres, and the markets those centres serve, accentuates the vulnerability of resource-producing regions to the vagaries of the market. Historically, that beaver felt hats and feather boas went out of fashion in Europe had catastrophic impacts on the Canadian fur trade in places such as Fort St James and on South Africa's ostrich industry in places such as Oudt-shoorn. Similarly, the 2014 global downturn in energy prices and demand had dramatic impacts in the oil-producing towns of Alberta.

A fourth basic concept often associated with descriptions of the capitalist economy involves the 'social relations of production'. This includes both the negotiated and the regulated forms and norms of behaviour and interaction between capital and labour. It should not come as a surprise that, over time, the form of these negotiated and regulated norms have changed as the relative bargaining power and expectations of both capital and labour have changed. The inclusion of attention to the social relations of production is a reminder that production is not simply a mechanical process, but that it also has a critical social dimension.

Prior to the 1980s global economic recession, labour prospered in many resource-producing rural, remote, and small town places and regions within developed economies. The 'long boom' of the post-war industrial expansion meant that the demands for raw material inputs remained generally high and there was as yet little competition in the supply of those raw materials from less-developed or developing regions. For labour, wages were generally quite high relative to national averages, strong unions helped to protect workers' rights and improve working conditions over time, and there was general job security as markets remained robust or were expanding.

After the 1980s, however, successive economic booms and recessions have shaken markets, international trade agreements have opened the way for raw materials from low-cost production nations to seamlessly enter the global supply chain, and capital has shifted its investments to take advantage of these new realities. In older resource-producing regions, the closure and downsizing of resource industry operations has reduced the bargaining power of labour. In many cases, capital simply has to close a mill or mine in order to force concessions from both labour and government. In BC's forest industry, once a bastion of labour power, unions have accepted dramatic rollbacks in wages and benefits as part of local agreements to reopen mills. The once powerful IWA union (International Woodworkers of America) was so reduced in members that it 'disappeared' through a merger with the United Steelworkers Union. In Finland, Halonen et al. (2017) highlight the dramatic employment losses that were experienced within the forestry and wood products sector. Their case study of Lieksa, in rural eastern Finland, demonstrates how these nationally notable employment losses were even more dramatic in smaller resource-dependent places. In New Zealand, Connelly and Nel (2017b) highlight how dramatic employment losses were experienced across the number of industries, including agriculture and mining. The impacts of these changes included not only increases in households dependent upon support from the state, but also shifts in employment shares towards services – another economic sector negatively impacted by economic restructuring and resultant downsizing or closure within non-metropolitan places and regions.

Finally, there is the notion of change and the historical trajectories embodied within past changes. As noted, historical trajectories have created patterns of dependence (on jobs and tax revenues) in the older and long-established resource-producing regions of developed economies, and has also created parallel vulnerabilities to the demands by global markets for local resources. This has established 'path dependence' in these resource-producing regions – a path dependence that is difficult to break away from. For example, in continuing the example from the preceding paragraph, not only has labour given away wage and benefits concessions to support the profitability needs of capital and keep local production going, but it has also joined with capital to lobby and put pressure on both local and senior levels of government to relax the tax or regulatory burden on the resource sector to keep it operating in these settings.

As will be examined later in the book, the historical trajectories that had bound capital and labour together in the resource-producing hinterlands of developed economies has created an uneven alliance through the more recent and intense period of resource industry restructuring. The current problem in that alliance is that while capital is mobile and able to invest in low-cost production regions, labour is relatively more fixed. Past examples, such as the Iron Ore Company of Canada taking profits from its Canadian mining activity to invest in lower-cost mines in South America that then replaced and closed the Canadian operations (Bradbury and St Martin, 1983) or current examples such as British Columbia-based West Fraser Timber's redirection of investment into lower operating-cost sawmills in the Southeastern US (where they now produce more lumber than their operations in Canada) (Halseth et al., 2017a) highlight some of this unevenness in mobility.

Nature of capitalism

In describing the nature of capitalism, Marx argued five basic and inter-connected points. The first, which we hinted at above, is to recognize that labour is a commodity. In a capitalist economy, workers exercise their capacity to work and they sell that capacity. As noted above, the restructuring of resource industries has placed a range of pressures on labour. First, there are fewer jobs per volume output in most sectors, and in many places this means simply fewer jobs in these industries full stop. Second, where jobs do remain in these industries, the bargaining power of labour has been reduced with the result that wages and benefits are also often reduced (sometimes relatively, sometimes absolutely). Third, as capital has relocated to lower-cost production regions, mills and mines closed with catastrophic local / regional job losses.

The second point involves recognition that there is a class-based control of production and production processes. In this system, capital purchases labour for a wage. As noted, over time the relative bargaining power of labour has been reduced as industrial efficiencies, the relative declining cost of transportation, international trade agreements that open markets to low-cost production regions, and the accelerated pace of change in an increasingly fast-paced global economy all contribute.

The third point, which connects the previous two, is recognition that there is an unequal exchange between capital and workers in the provision / purchase of labour. As noted, capital will seek to increase the difference between the value of labour and the value of labour power in order to enhance profitability. That is, capital not only sets wages and organizes production, but it makes choices about when and where it is most profitable to locate the different labour processes within the production chain. Labour, by contrast, is less mobile and thus has more relatively limited bargaining power unless specific rights are designated by the state. Even in that case, capital has the ultimate bargaining tool – its capacity to close production and eliminate employment opportunities for workers.

In most Marxist analyses, the unequal exchange between capital and labour is considered an exploitive relationship. That exploitation derives in large measure from the focus of the capitalist system on maximizing profitability and return on investment. Again, this involves maximizing the difference between the value of labour and the value of labour power. Therefore, the fourth point important in this description of the nature of capitalism is to recognize profit as the dynamic that drives the system.

In older resource-production regions, there is often long experience with the complexities of the profitability imperative for capital. Prior to the 1980s, labour was able to make use of the profitability imperative by threatening to withhold labour (through strikes or similar job action). Faced with the potential for significant disruptions in production (and thus income and profit), capital would generally agree to higher wages and enhanced benefits – both of which were considered small costs relative to the possibility of continued profitability. While this has reversed in some sectors (such as noted above for forestry) it continues in other sectors (such as oil and gas) where the profitability potential generally remains very high and thus the wage / benefit package for individual workers remains relatively lucrative. After the 1980s, the substitution of automated production processes was greeted as a mixed blessing in these older resource-producing regions. On the one hand, the increased efficiency (and thus profitability) of the operation guaranteed that (at least for the short term) the industry would stay in operation. However, these were labour-saving technologies and processes, and each new investment meant the loss of jobs, together with all the associated local economic spinoffs, and it also meant the likely out-migration of families from remote, rural, or small town places. Finally, it also meant that drives for local profitability may have only undercut future sustainability in the industry through threats to the long-term supply of both renewable and non-renewable resources, or through the reinvestment of local profits into facilities in low-cost production regions.

The fifth point recognizes that the capitalist economy is generally antagonistic towards social development. In other words, it is not within the interests of capital to invest in the needs of labour beyond that which is essential to reproducing labour for participation in the production process.

The way these basic concepts, and the five points which describe the nature of capitalism, come together at any particular time and in any particular place will define the particular form of the production process and the particular nature of the capital–labour relationship. Together, these arrangements are called 'modes of production'. Such modes of production will, of course, not only be different and will vary from place to place, but they will also vary from sector to sector. Over time and across places, the way in which these modes of production change will also vary. This further supports the processes of spatial differentiation and uneven development already introduced.

Restructuring

Marx argued that, at any given time, relations within the capitalist mode of production were historically produced. Like with other notions and perspectives discussed thus far, this means that explanations are contingent on the trajectories and legacies of past economies, societies, and even specific events. Naturally, changes will occur over time to a mode of production as a result of technical, political, or other transformations. In the paragraphs below, we pay specific attention to the shift from a 'Fordist' mode of production to a 'Flexible' mode of production, including reviewing some of the key differences between the two. It is important to note at the outset that many labels have been employed in describing the transition to more flexible systems of production. In Canada's forestry sector, for example, Hayter (2000) links to the literatures on flexible specialization, flexible accumulation, post-Fordism, and the information and communication based 'techno-economic' paradigm. Before we get to a discussion comparing Fordist and Flexible modes or production, however, we want to discuss the general topic of change.

Change is, of course, not a new process, whether it is in our daily lives or in the broader structures organizing the global economy. As well, changes in the structure and organization of the global economy are ongoing. Sometimes, however, the heat of the moment around sudden changes can fool us into believing that change had previously been either not important or not present. As the rural sociologist Janet Fitchen (1991, p. 259) reminded us, it is only by being immersed in change that "makes the past appear stable and unchanging by contrast". If we were to look back at the mobility of capital, labour, raw materials, and final products, there is little that separates the beaver felt hat from the modern smartphone (we pick the beaver felt hat because of the importance of the beaver for trade in the historical development of Canada) – international capital, seeking profit for shareholders, organizes a global supply chain that begins with access to the raw material inputs and moves all the way through to advanced manufacturing centred upon a particular luxury product that could, in the market of the day, be sold at a significant profit. Time and effort is also expended by capital to 'grow' the market for the product and to nurture consumer tastes and demands.

While the comparison of the beaver felt hat and the modern smartphone seems ridiculous, it does serve to illustrate Harvey's (1990) critique of restructuring – what exactly is it that is changing? The fact that we see changes to the form and nature of work in resource-production regions, and that theorists push us to distinguish between the observable (or the 'concrete') and the underlying structure (or the 'abstract') has meant that discussion of restructuring is replete in the literature, and that it is also both contested and problematic.

In setting up his discussion of the transition between Fordism and Flexible production, Harvey (1990, p. 124) acknowledges that

> the long postwar boom, from 1945 to 1973, was built upon a certain set of labour control practices, technological mixes, consumption habits, and configurations of political-economic power, and that this configuration can reasonably be called Fordist-Keynesian. The breakup of this system since 1973 has inaugurated a period of rapid change, flux, and uncertainty.

In seeking to differentiate the concrete from the abstract, Harvey (1990, p. 124) warns: "There is always a danger of confusing the transitory and the ephemeral with more fundamental transformations in political-economic life".

To start to answer questions about change, let's start with the 'concrete' – the observable manifestations that have occupied the restructuring debate. In trying to organize a description of the characteristics of a Fordist mode of production relative to a Flexible mode of production, we want to draw out and highlight four topic areas. Drawing upon Harvey (1990), who in turns draws upon Swyngedouw (1986), these four topic areas include the organiza-tion of work, the organization of production, the scale of production, and the reproduction of labour.

The organization of work identifies the ways by which labour is brought into the production process. Under a Fordist mode of production, workers were given individual tasks along a simplified production line. Relative to earlier craft industries, this was considered a critical 'deskilling' of the work-force. For capital, the opportunity would be to switch out workers relatively easily, since the training needed for any given task was minimal. For labour, not only was there deskilling, but the new style of work and worker were confined to single and repetitive tasks. In addition, the pace of the labour process now moved at the pace of the assembly line. In a Flexible mode of production, the emphasis is upon 'multi-tasking' and the creation of a team environment that looks at the labouring component of the production process as units rather than as single workers along an assembly line.

A good deal of the industrial conflict in resource-producing regions from the early 1980s through to the 2010s focused upon this specific aspect of restructuring. For capital, this not only included an attempt to introduce greater flexibility into the production process by having workers and their equipment do more than one task, it also was seen as a significant way towards a numerically smaller workforce and a less complicated labour contract that had fewer job categories. For labour, it was clear that 'flexibility' was simply a euphemism for fewer workers. With the waning of labour's bargaining power in many resource industry sectors in developed economies, the adoption of flexible work arrangements has become relatively widespread.

There was also considerable transformation in the organization of production. Under a Fordist mode of production, there was a clear separation of the factory or 'shop' floor from the design, marketing, and management functions of the production process. This served to create clear and rigid hierarchies within the workplace. The removal of workers from the design and management elements (that have been common in craft industries), meant that adjustments or improvements discovered during the production of a product could not be implemented because there were few, if any, avenues for dialogue.

Under a Flexible mode of production, there is a specific attempt to break down the barriers between design, management, and production. Effectively, workers were encouraged to participate in quality control and product improvement dialogue. Combined with the development of a team approach, there was an attempt by capital to co-opt labour into the capitalist agenda by showing that all of their interests were served by creating better-designed products that were produced in smarter ways, and thus would be more competitive in the marketplace. The rigid job classifications and hierarchies were replaced with production teams, or as some industries began to call them, production platforms.

These first two transformations had a significant impact in older resource-producing regions within developed economies. In addition to the loss of jobs, and the co-option of labour through a production team / platform approach, there is also the question of outsourcing. Flexible production became married with concerns about obtaining lower-cost inputs into specific parts of the production process. This meant the increasing use of third-party or offshore production.

The third significant transformation that is widely noted in the literature involves the scale of production. Building off of its automobile manufacturing stereotype, a Fordist mode of production emphasized very large production runs with very limited product differentiation. This echoes the legendary advertising slogan for Ford's early Model T automobile: 'you can have any color you want, as long as that color is black'. These large-scale production runs were well suited to assembly-line manufacturing and the isolated shop floor environment – which, once set up, could run for long periods of time with relatively little adjustment or oversight.

A Fordist mode of production was particularly well-suited to industrial resource exploitation. Whether in minerals, forest products, or agricultural commodities, large production runs of a relatively undifferentiated product were the norm. It also fit well with the low value-added, high-volume approach to production that was needed to generate sufficient profit and return on capital investment in these industries.

By the 1980s, large production runs of similar products encountered two critical problems. The first was that once people had purchased a product, the opportunity to generate additional revenue from product sales to those consumers was over. One had to wait until the product wore out, broke, or was used up. To continue selling, there was an imperative to get consumers to wish to purchase another product as soon as possible. For many products, this meant accelerating the obsolescence curve. Today, the annual rate with which people replace their mobile phones differs fantastically from the 1950s and 1960s, when a household might never replace its landline telephone or might upgrade once every decade as something radically new such as push buttons or cordless options came into vogue. The second was simply related to the increasing pace of the global economy and the rise of alternative products. Large production runs for most consumer products sectors were no longer viable.

Under a Flexible mode of production, the emphasis is on small batches and limited production runs. This includes opportunities to introduce specialized products and niche products. In some cases, products can be specialized to individual consumer specifications. In resource industries, these questions about the scale of production and the development of niche products is complicated. At one level, authors such as Troughton (2005) have argued that

Fordist-style production essentially continues to be the dominant mode. Large volumes of a relatively undifferentiated product continue to be moved in alignment with older profitability imperatives and the continuing structure of both the industry and the marketplace. Others have looked to identify ways by which the industry has become more flexible and capable of specialized production runs. This includes 'mixing' different forms or grades of a raw material in order to meet specific consumer requirements. Coal from different mines is now combined in different volumes so as to reach a certain BTU value for a specific application or customer. Similarly, different grain products may be combined in order to reach different specified caloric values for different customers. The application of computer instrumentation and further auto-mation in the pulp and paper industry has made that industry much more flexible, with some mills having been converted to speciality mills that focus on small batch production tailored to the specifications and needs of different customers. The additional costs of such flexible production are often com-pensated for by securing additional market share or by charging a premium for the product.

The final category often used to describe some of the outcomes of restruc-turing focus upon the reproduction of labour. As described earlier, under a Fordist mode of production, capital purchases labour for a wage. In many cases, wages were tied closely to the profitability of the sector. Labour relations under this mode of production were generally confrontational, with both sides using closure / withdrawal of work as their levers to win concessions from the other side.

Under a Flexible mode of production, the reproduction of labour now seeks to encourage workers to 'buy into' the corporate family. The previously con-frontational approach to labour relations and collective bargaining has been replaced with attempts to develop partnerships. Such partnerships may include profit-sharing schemes and labour (or even individual labourers) taking up ownership shares in the company. In addition, a wage-focused settlement package has now been replaced with ones focused on a wider range of social benefits. This even includes negotiations such that corporate pension plans may be encouraged to put their money into 'green' or sustainable investments.

Restructuring – or not?

Having described some of the differences between the Fordist mode of pro-duction and a more Flexible mode of production, we can now talk about the process of transition between the two – or the process of restructuring. As described by Hayter (2000, p. 11), by the mid-1970s "Fordism was increas-ingly threatened as a system of regulation and as a set of productive arrangements ... [these] arrangements were undermined by technological change, market dynamics, and recession".

The process of transition would be difficult in resource-dependent regions with a narrow economic base. One of the central features of restructuring has been the transformation of labour and labour relations. However, the shift from a Fordist to Flexible workforce has not only been contested, but the very definitions that frame the transition are problematic. As Hayter (2000, p. 255) argues, under a Fordist labour arrangement, the workplace was:

> based on the Taylorist principles of seniority and job demarcation ... [However,] ... the search for more flexible workforces is a contested and ambiguous process. The search is contested because the demands for worker flexibility are usually part of employment downsizing and require changes in established, legally sanctioned agreements. The search is ambiguous because the concept of employment flexibility is multifaceted and no consensus has been reached on an appropriate definition. Thus, flexibility may be realized through a low-wage, low-skilled, unstable workforce, or buy a high-wage, multiskilled, stable workforce, or by other combinations of characteristics.

At the concrete level, there is really no question that there have been changes in the production process via the restructuring from a Fordist to a Flexible mode of production. But, transitioning, or restructuring, within the global economy is at any given time complex, incomplete, and uneven. The dominant post-Second World War structure of Fordist-style assembly-line mass production has given way in many sectors to a more Flexible mode of production. Despite this perceived restructuring, many forms of the older Fordist style production remain.

Against such trends, or at least perceived trends, are the ways capital selects and mixes choices and elements for both production and consumption, and then distributes these globally. While some may have given Fordist mass production its final send-off, the rise of China as the 'workshop of the world' looks a great deal like Fordist mass production, aided in part by technologies that allow for some degree of production flexibility. Similarly, the rise of the information economy has seen a massive growth in online services – only to have those services concentrate in low-cost production countries such as India, where Fordist-style factory floors churn out tech advice given by a highly segmented labour force working through rote sets of instructions before pushing the client on to the next stop in the information / production assembly line.

The answer to the question about whether there has been a fundamental restructuring of the capitalist political economy is maybe even more unclear once the analysis turns to the 'abstract'. In this, the question becomes whether the underlying imperatives and structures of capitalism have been transformed. At this level of analysis, basic concepts discussed above, together with the five elements discussed under the nature of capitalism, appear to not only remain in place, but to have been reinforced. At its core, Harvey (1990, p. 121) argues that "the basic rules of a capitalist mode of production continue to operate as invariant shaping forces in historical-geographical development". As noted above, there have been changes in the mode of production, but not to the underlying constructs and imperatives.

In terms of the nature of capitalism, labour very much remains a commodity that is increasingly traded. In some cases, labour is moved, while in most cases, it is capital which relocates to places of increased labour advantage. As a result, it is clear that there remains a class-based control of production and that the unequal exchange between capital and labour is perpetuated within that class-based control. Profit remains the dynamic that drives the system, and the search for profit amid successive crises of capitalism demonstrates a continuing antagonism to social development. Following his own detailed critique, Harvey (1990, p. 189) concludes "that there has certainly been a sea-change in the surface appearance of capitalism since 1973, even though the underlying logic of capitalist accumulation and its crisis-tendencies remain the same".

If the fundamental needs and underlying logic of capital remain unchanged, what then has changed? Harvey (2005, p. 4) argues that the processes of 'space-time compression' have been fundamental to the acceleration of change within the concrete expression of capitalism: "technologies have compressed the rising density of market transactions in both space and time. They have produced a particularly intensive burst of what I have elsewhere called 'time–space compression'." Change is not only a long-term and ongoing process, but through the advent of new communications and transportation technologies, it is an accelerating process. Looking forward, Harvey (1990, p. 188) argues that as Flexible accumulation emerges, it will itself transition and will be recognized as a transitory form of the capitalist political economy, because, in fact, "capitalism is a constantly revolutionary force in world history, a force that perpetually re-shapes the world into new and often quite unexpected configurations".

The implications of restructuring for resource-dependent regions and places are significant. Beginning with the concrete, there have been tremendous changes in the organization and management of labour. At mines where up to 1,200 workers used to be employed, flexible job categories and the aggressive application of computer-based technologies have shed jobs such that these mines now function with only one-third to one-half of their workforce of just 20 years earlier. At one mine we are familiar with, a blasting crew of 24 workers was replaced with an automated, computer-controlled, GPS-positioned 'robot', operated by one person, coordinated by a second person, and maintained by a third person.

In addition to the job losses just described, the contracting-out of work has also cut the permanent workforce. Contracting out work previously done by company workers to 'independent contractors' creates a vulnerable cohort of entrepreneurs, whose services and compensation rates are almost always entirely linked to the local resource industry. Who gets hired, when, for how long, and for what rates makes independent contractors far more vulnerable to exploitation than unionized company employees. Should the large resource industry close, or declare bankruptcy, these independent contractors are generally not on the recognized creditor lists (banks, in contrast, are generally always listed first or second) and can, therefore, lose any chance of recouping what might be owed to them.

Labour is also being transformed by the discourses of profitability. As described by Halseth et al. (2017b), labour has bought into the capital imperative for required high levels of profitability to the degree where unions have granted concessions around wage levels and benefits as part of efforts to get older mills or mines up and running again. The dialogue around partnerships and teamwork that has underwritten much of this discourse, and the actions by organized labour, only rarely is extended into formal profit sharing arrangements so that both capital and labour benefit equally.

Among the most dramatic of impacts, however, relate to the implications of time–space compression on resource industry operations. With profitability linked closely to commodity prices, global affairs and events can impact commodity prices almost overnight. When commodity prices fall below production costs, the potential for profits will change over into the potential for operating losses, with the result that today's connected and increasingly automated mills and mines can be shut down with almost no notice. Drawing again from our experience in northern BC, we know of cases where the mill manager only found out about the closure of his operation as he arrived at work in the morning to see the doors locked and machines already shutting down. Resource industries in developed economies, because they have focused on efficiency of operations as part of a drive to compete globally, have become de facto 'turn key' operations – if commodity prices rise above a specified level, the mill / mine can be turned on in a few days, and if commodity prices fall below a specified level, the mill / mine can often be closed in the single work shift.

Closing

The purpose of this chapter has been to provide an introduction to the basic form and characteristics of the capitalist economy. As the readers no doubt have gathered, the literatures, and the debates within and between the literatures, on the nature of capitalism are big and often not as polite to one another as we have been in combining their insights. While we have tried to stay away from the dense and impenetrable language that some theories seem to enjoy, it has nonetheless been complicated. However, this is a critical building block in developing our understanding of resource-dependent places and economies so as to better interpret how those places and economies are changing.

The first part of the chapter employed a Marxist analysis to describe the nature of capitalism. Especially important was the identification of capitalism not so much as a thing, nor a mechanical set of economic transactions, but rather as a complex and dynamic social process. This social process perspective helps inform our construction of a political economy of resource-dependent places. Another important aspect was the delineation of the concrete from the abstract. This proved especially helpful in preventing us from being overly influenced by the transitory modes of capitalism and to stay focused on its underlying imperatives. These imperatives include the need for capital to organize access to raw materials, to manage material production, and to control labour so as to energize that production process. It also includes that profit is the key dynamic that drives the system and that this brings together both production and consumption processes.

Since the 1980s, dynamic global changes in the capitalist economy have created a series of crises in capitalism such that social and political power has been marshalled to support the increasing globalization of trade and economic flows. The opportunity for capital to access historic markets while relocating production to low-cost states explains some of the closures seen in remote, rural, and small town resource places and regions in developed economies.

In assessing changes in the economy, we compared and contrasted the Fordist mode of production, which dominated the capitalist economic system within developed economies from the 1940s to the 1980s, with the emergent Flexible mode of production. Across a number of elements, it was clear that the observable mode of production had shifted and changed as a result of recurrent crises of profitability and the opportunities being created by the time–space compressing potential of new information technologies and international trade agreements.

In critiquing that assessment of the changes between the Fordist and Flexible modes of production, we noted that the underlying structures of capitalism had remained unchanged. The implications for resource-producing places and regions have been significant. At one level, there have been job losses and new forms of labour vulnerability have been created. At another level, local resource industry operations now function on not just a leaner, but also a more contingent, basis. Depending upon commodity prices, resource-processing operations are now much easier to open and close in order to protect profitability and the corporation's economic bottom line. At still another level, the options to relocate to low-cost jurisdictions has meant the closure of production facilities in developed economies as transnational corporations seek locations with lower wages, fewer regulatory burdens, limited environmental regulation, and the like.

This connectivity of the capitalist system, of the connected relationship between resource-producing regions and transnational corporations, and of the connections between operations now located in multiple countries, highlights and reinforces not only that capitalism is a social process, but also that is a global system. The next chapter turns to consideration of the nature of that global capitalist system.

4 Global capitalist system

Introduction

As described in Chapter 3, we live and work in a capitalist economy. This means that processes of production and consumption are linked via markets. It also means that within this broader organizing structure, capital and labour are together engaged in an ongoing debate and negotiation about the particular form of production that can be found in different places at different times. These are political and social processes, as well as economic, and they occur both in places and across spaces.

Resource-producing places and regions in developed economies constitute particular constructions of this relationship between production and consumption and, as we shall find, their 'geography' is complex. For example, they often coalesce into regional concentrations based on the availability of specific raw materials and access to the needed types of infrastructure, services, expertise, energy, and labour. These resource-producing regions are often peripheral to the political, economic, decision-making, and population centres of their respective states. And these resource-producing areas are also peripheral to the global economy as they are suppliers of raw materials that serve as inputs for external manufacturing centres and regions.

These 'geographies' necessitate consideration of the wider system within which resource-producing regions function. As a result, this chapter looks at elements of the spatial organizing structures within the global capitalist system. To start, the chapter introduces the notion of globalization. While the term has received considerable press over the last decade, the globalization of the economy is nothing new – what may be new are the transportation and communications technologies that have accelerated the pace of the global economy, as well as the processes of international agreements that now facilitate the freer movement of capital and goods. To understand the structural relationships within this global system, we introduce World Systems Analysis. This approach extends our consideration of the capitalist economy and establishes the basis for understanding relations between peripheral (or dependent) regions and core regions within the global economy. Notions of dependency and marginality within the global economic system are especially important for resource-producing places and regions. The chapter then looks at geographic models describing core–periphery, or heartland–hinterland, type relationships in order to incorporate consideration of the particular attributes of resource-producing regions within developed economies.

Globalization

As noted, this chapter builds on the previous discussion of the capitalist economy and adds an overt interest in the changing spatial relations of that economy. As we tell our students who have been fans of those various pirate movies, both colonial empires and mercantilist trading alliances were global in reach. So too were companies such as the Hudson's Bay Company and the Dutch East India Company that flourished within those political-economic systems. Today, it takes no stretch of the imagination to know that the capitalist economy is also globally organized. A quick look at the news or the packaging of the various food or consumer products we might buy reports to us that transnational firms organize their inputs of materials and labour, and manage their production and distribution operations, not only globally but also in real time. They also report to us stories of how transnational firms will pursue cost-price advantage by locating and re-locating aspects of their production and marketing chains as global circumstances change. Drawing upon David Harvey and Neil Smith, Barnes (1996) describes how the changing relations between capital and labour across space leads to an ongoing process of investments and disinvestments in places leading to uneven development.

But globalization involves more than simply 'cross-border' or 'trans-border' trade or exchange. Globalization is important for the ways that it is changing economic and political relations on a global scale. To explore its implications for resource-dependent remote, rural, and small town places in developed economies, we need to look into this concept. To do this, the following is broken down into two units. The first provides an introduction to the concept of 'globalization' while the second introduces the concept of the 'new international division of labour'.

The concept of globalization

Ritzer and Dean (2015) present a very useful textbook treatment of the concept of globalization. For them, globalization involves sets of processes that support faster and more multi-directional flows of people, goods, services, and information around the world. Far more reaching in scope and simply 'transnationalism', the sets of processes embodied in globalization require various physical, institutional, and regulatory structures through which they may operate. These sets of processes themselves also play a significant role in creating, modifying, and even destroying, such physical, institutional, and regulatory structures. The scope of globalization is explored in their text via topics that range from economic flows to the flows of people, from cultural to environmental issues, and from policy to political structures. They also raise topics connected to uneven development. In their words, they highlight 'inequalities' at global, regional (including urban-rural), and local levels.

Stutz and de Souza (1998) add the element of 'control' or domination as an important part of the concept, and the challenge, of understanding the world economy. They argue that

> an understanding of the reasons for problems in the world economy begins by recognizing its domination by developed countries and the existence of an international economic order established as a framework for an international economic system. The term *world economy* refers to the capitalist world economy, a multistate economic system that was created in the late 15th and early 16th centuries. As the system expanded, it took on the configuration of a core of dominant countries with the periphery of dominated countries.
>
> (Stutz and de Souza, 1998, p. 25; italics in original)

They go on to identify that the global economy is "an evolving market system in which a hierarchy of states develop based on economic development. The labels core, periphery, and semi-periphery are used to identify economic processes that operate at different levels of this world economy hierarchy" (Stutz and de Souza, 1998, p. 25). In the next section, we recognize and explore the ongoing negotiation and re-negotiation of positions and power that occurs between countries of the globalized core and those of the globalized periphery.

The concept of globalization is complex and, therefore, difficult to define. The Organisation for Economic Co-operation and Development (OECD) provided an early conceptual definition of globalization as the

> increased movement of tangible and intangible goods and services, including ownership rights, via trade and investment, and often of people, via migration. It can be, and often is, facilitated by a lowering of government impediments to the movement, and/or by technological progress, notably in transportation and communications.
>
> (Oman, 1996, p. 6)

The increased movement of tangible goods and services also enhances the connections and thus the dependencies between places and regions within the trading system. Cohn (2000, p. 10) describes globalization as "a process that has two major aspects: the broadening and the deepening of interactions and interdependence among societies and states throughout the world". Linking to discussions in the previous chapter about how the contradictions and tensions within capitalism create the foundations for uneven investment and development in both social and geographic scales, Cohn (2000, p. 10) reinforces that even "globalization is not a uniform process throughout the world".

Stutz and de Souza (1998) identify three important dimensions of globalization, including the globalization of culture, the globalization of the economy, and the globalization of environmental change. With respect to the globalization of the economy, they highlight aspects including the globalization of consumption, the role that increasingly fast global telecommunications plays in communicating norms around production and consumption but which also facilitate 'real-time' economic management. Interconnectivity is key.

Stutz and de Souza (1998) also identify seven key trends with respect to the globalization of the economy. These include:

- the globalization of finance,
- the increasing importance of transnational corporations,
- global direct foreign investment from core regions into less developed regions,
- global specialization in the location of production,
- globalization of the tertiary sector,
- globalization of the office function, and
- global tourism.

The globalization of finance is linked to two transformations. The first is the role of supra-national trading and economic blocs, as well as new international trade agreements that facilitate investment and investment protection (Jessop, 2013). The second is the role of modern telecommunications, and computer-managed financial monitoring and investment decision making, which has revolutionized global finance, not necessarily through function, but through the pace and speed of transactions. With different markets, opening at different times, all around the world, it seems that 'capital never sleeps'.

International trade agreements that protect investment and 'open' markets are not a new phenomenon. Under the British Empire, for example, various forms of 'preferential trade' agreements were set in place to support colonies in getting past the significant trade barriers and tariffs that ringed Great Britain's home market. What is new, however, is that they now come with their own governance and regulatory powers that either strip away jurisdiction or authority from states, constrain the actions that states may take to manage their own internal economies, and they allow capital to take legal recourse if it feels aggrieved by state decisions or policy changes. Sparke (2002, p. 221), quoting Neil Smith's work on capital and power, expands on these processes where corporations take advantage of new trade pathways and their implications of re-scaled global economic organization on our perception of place to argue,

> Capital, the guardians of information flow, information corporations – 'the power holding organizations' – may entertain the fantasy of spacelessness and act accordingly, but in practice, every strategy to avoid and

supercede 'historically established mechanisms' and territories of social control involves not the extinction of place per se but the reinvention of place at a different scale – a capital-centred jumping of scale.

(Smith, 1996, p. 72)

Examples of international trade agreements include bi-lateral or multi-lateral agreements, such as the North American Free Trade Agreement (NAFTA) or the Trans-Pacific Partnership (TPP). NAFTA originally included two (in the Canada–United States Free Trade Agreement – FTA), and later expanded to three (NAFTA) states, while the Trans-Pacific Partnership was originally negotiated with 12 participating states. Such trade agreements have a spatial organization or spatial logic underlying the partnership. Often this includes access to manufacturing markets for the raw materials from resource-producing states and access to populous markets for manufacturing states (as well as access to cheaper raw materials for their manufacturing sector). As noted above, there is also a political element with states giving up some of their sovereignty in order to 'ease' the flow of people, money, materials, goods, or services within the trading bloc. Such agreements may also represent a form of 'globalization in miniature'.

Supporting the increasingly globalized economy are a range of institutions that smooth the processes of market exchange and provide reference points for the expansion of capital. The management side of globalization includes, therefore, a host of world economic organizations – many of which were first created in the immediate post-war period as states sought to curb the excesses of capital that had led to the Great Depression and the unrest that supported the rise of nationalist governments in Italy and Germany. As noted by Harvey (1990, p. 137) one of the first was the Bretton Woods Agreement in 1944, which "turned the [US] dollar into the world's reserve currency and tied the world's economic development firmly to US fiscal and monetary policy". Other institutions were also established partly in response to the emerging Cold War. Examples of such world economic organizations are the International Monetary Fund (IMF) and the World Bank (first set up as the International Bank for Reconstruction and Development).

As Corbridge (1994, p. 290) notes, "In this new world, power is rapidly being devolved to the markets and to the IMF and the World Bank". The impacts for resource-producing regions are twofold. For such regions in developed economies, there is increased competition from low-cost regions of the world. This puts downward pressure on profitability, wages, and investments in these locations. Such competition is behind the closure of many resource industries and the relocation of capital from developed economies to low-cost regions. Within those low-cost regions, there have also been various economic transition funding supports provided by international organizations that each come with rules whereby these low-cost producing regions "are being forced to export more primary commodities as part of their agreements with the IMF and the World Bank" (Corbridge, 1994, p. 292). Globalization, in other words, has created a circumstance whereby resource production in developed economies is struggling with the flow of lower-cost commodity production, while developing economies seeking to 'move up' the production value chain are instead being forced to put even more of their low-cost commodities into the global marketplace – something which puts further downward pressure on international commodity prices.

The style of these global economic institutions was modelled after the developed economies of the day, and the application of their policies is part of a wider strategy to engage other countries in the global capitalist marketplace. The active intervention into national economies by these institutions of globalization effectively 'hooked' countries of the globalized periphery through loans and debt repayment schedules where "stringent limits were placed on the economic policies of debtors, with the result that a majority of citizens in these nations often found themselves worse off" (Stutz and de Souza, 1998, p. 31). Loans and debt remain some of the most effective tools to ensure that countries of the globalized periphery participate in the capitalist economy and remain open to foreign direct investment.

There are also some world economic processes that have become de facto world economic organizations. These would include the General Agreement on Trade and Tariffs (GATT), which through successive 'rounds' of negotiations have set international rules and protocols for the management of trade and trade-related investment.

Within this changing world of institutions, the increasing importance of transnational corporations is due to the fact that they are now the primary agents of international trade. Stutz and de Souza (1998, p. 10) highlight that "the globalization of the economy has been spearheaded by transnational corporations" which "may conduct research, operate industries, and sell products in many countries, not just where its headquarters are based". They manage their access to raw materials, to the various factors of production (such as land, labour, finance, energy, etc.), and to markets according to the characteristics (tax regimes, regulations, costs, etc.) of different countries.

Facilitated by the globalization of finance and computer-based communications of management, production, and consumption, transnational corporations will use the 'geographies of difference' between countries in order to best manage their different operations for profit. For example, all things being equal, they might shift the location of their most labour-intensive manufacturing operations to states which have low labour costs, or they may choose to allocate their transferable "taxable profits to the lowest tax country possible and [thus] minimize tax overhead" (Stutz and de Souza, 1998, p. 11). As seen in the laptop computer example above, the entire world functions as the 'shop floor' for transnational corporations.

New international division of labour

The increasingly global distribution of corporate activities is also part of the 'new international division of labour' that emerged after the global economic crises of the early 1980s. The bow wave of this new international division of labour saw the closure of manufacturing in high-wage high-regulation counties of the globalized core and the transfer of these activities to the low-wage and low-regulation countries of the globalized periphery. Mexico, East Asia, and South Asia were among the early recipients of this re-direction of manufacturing investment. From textiles to consumer goods to microchips, the pattern seems to hold that once new products find their place in the market their production begins to shift to lower and lower production-cost countries.

Under ever-accelerating pressures of competition and the requirement for continued profitability, capital continues its restless search for global cost advantages. In 2014 and 2015, consumer electronics giant Samsung announced that it would be opening a new US$3 billion plant in Vietnam to house part of its smartphone manufacturing. With rising prosperity in China, wages in that country are no longer the lowest in the region. To create a product that is still price competitive with other smartphone companies "Samsung is joining Microsoft Corp., LG Electronics Inc. and others in decamping to Vietnam where tax breaks are plenty and wages are lower" (International Business Times, 2014). In targeting the emerging smartphone markets in India and China,

> Samsung needs to be a lot more price competitive particularly at the lower end, and driving down manufacturing costs by moving to Vietnam will help. The GDP per capita in Vietnam is about a quarter than that of China, so Samsung would be able to "reduce their labor bill by two or three times probably".
>
> (International Business Times, 2014)

The outsourcing of raw material supplies, communication services, or manu-facturing operations has led to increased foreign direct investment in nations across the globalized periphery. For such nations, this is export-led indus-trialization and it is "characterized by countries welcoming foreign investment to build factories that will manufacture goods for international markets and employ local labor. Export-led policies rely on global capital markets to facilitate international investment and global marketing networks to distribute the products" (Stutz and de Souza, 1998, p. 16). To participate, countries across the globalized periphery must compete with one another for direct foreign investments. With a number of competing countries across the globa-lized periphery eager for such investments, there is an uneven bargaining advantage to transnational corporations.

As suggested above, the process is a 'globalization via foreign direct investment', and the dispersal of corporate functions into countries with the 'right mix' of cost advantages has impacted all stages of business from finance to research, logistics, production, and consumption. Increasing attention in recent years has been directed to the globalization of the tertiary sector. This includes a vast array of services including legal, financial, consulting, advertising and marketing, educational, accounting and billing, computer technology services, and many other 'back room' office and clerical functions. Another area receiving press is the mobility of the 'online' customer service or ordering functions. Using the example of south Asia, Stutz and de Souza (1998, p. 19) describe how the "government of India provided all the requisite inputs for transnationals to hire local talent – training for manpower, high-speed data communications, red-tape-free systems, and a virtual red carpet" for foreign direct investment.

World Systems Analysis

While the preceding section highlighted how the 'global' organization of our economy is certainly not a new phenomenon, it also highlighted how its recent evolution now employs international trade agreements and law to enhance the mobility of capital. Against that backdrop of increasing advantages to the mobility of capital, we need to consider questions about the spatial patterns of investment and development. We need to consider the contemporary outcomes of dependency and why underdevelopment persists. Why do some lagging regions continue to lag despite massive growth in the global economy over the past 50 years? Why do most dominant regions remain dominant? In this section, we introduce some of the historically competing theories around 'development' and then focus on World Systems Analysis to draw insights about dependence and persistent underdevelopment.

Modernization theories

In general terms, two theoretical approaches have sought to explain or predict the 'pathways' to development. One of these approaches included what have been labelled as 'modernization' theories. Using the 'model' or 'exemplar' of an advanced modern economy, the premise was that different states or regions were simply at different points along a timeline or 'inevitable' development pathway to the same economic development destination.

Rostow's model of economic growth is one of the most cited of these modernization theories. Writing in the mid- to late-1950s, Rostow posited a model that involved a five-stage progression from a subsistence agricultural economy to an advanced industrial economy. These stages included: "traditional society', "preconditions for take-off", "take-off to sustained growth", "drive to maturity", and "age of high mass consumption" (Norton, 2004). The stages were generalizations of the progression witnessed in existing advanced economies.

The "stages" of Rostow's modernization model are relatively straightforward. Traditional society was one dependent upon subsistence agriculture and where there was relatively limited economic trade. The preconditions for take-off stage centres upon an external influence (such as colonialism or investment by transnational corporations) which create, or focus upon, opportunities to mobilize a local resource into the global economy. The take-off to sustained growth stage focuses upon the extensive exploitation of a major natural resource and the necessary social, political, and economic realignments of the region that will support the high-volume export of that resource. The drive to maturity stage includes the development of a diverse industrial base to take advantage of up-stream and down-stream economic and industrial development opportunities from that original resource exporting base. The age of high mass consumption is characterized by the emergence of a tertiary sector service and management economy that has grown significantly relative to the original resource-dependent economic base.

By creating a 'timeline' of sequential growth that was unidirectional towards an advanced industrial economy, the model assumed that there were no structural barriers within the global economy. Challenges to the Rostow model are, however, considerable. To start, later chapters will review the Staples thesis critique, which highlights and reinforces the emergence of dependency following exploitation of a major resource and identifies why a diversified economy (either up- or down-stream of the resource production base) does not occur. Additionally, the characterization of a capitalist economy in the preceding chapter demands patterns of uneven development so that accumulation can occur in select locations. The assumption that there will be a general equalization of levels of development would not create the necessary requisites for such accumulation processes. From the standpoint of either a Staples or Marxist critique, the inevitable pathway through a series of stages to an advanced economy is by no means inevitable. As most of the authors cited in this chapter also show, the evidence 'on the ground' of a general equalization in development is also lacking.

World Systems Analysis insights

A second body of theories focus on the global organization of the economy and seek to explain development pathways that focus on the persistence of inequality and the failure of underdeveloped regions to 'advance' to the same level of economic and social development as existing advanced industrial societies. Researchers have suggested that globalization, and the inter-dependencies created through such processes, have masked a new form of neocolonialism (Makuwira, 2007).

In their textbook on the global economy, Stutz and de Souza (1998) provide an excellent thumbnail of 'World Systems Analysis'. To start, they describe how it provides a "framework for understanding the development of the capitalist system and its three components – the core, the semiperiphery, and the periphery" (Stutz and de Souza, 1998, p. 545). They also highlight how this framework draws upon earlier work by André Gunder Frank, whose studies of the Latin American experience highlighted how "the development of the West depended upon the impoverishment of the periphery" (Stutz and de Souza, 1998, p. 545). As a counter to modernization theories, such underdevelopment theories pay specific attention to the issue of dependence (and thus vulnerability).

World Systems Analysis is also of critical assistance in the study of the changing political economy of resource-dependent regions, because it focuses upon inequality and, more importantly, the ongoing construction and re-construction of unequal relations. For Wallerstein, the "capitalist world-economy is necessarily unequal, comprised of peripheral processes (low-income and low-profit production) and core processes (high-income and high-profit). These processes cluster geographically, so that core processes may dominate" (Flint, 2009, p. 814). Cater and Jones (1989) also explored questions of dependence at regional and global scales. They described how the origins of dependency theory sought to illuminate the exploitive relationships between core and peripheral regions in the global economy. For them, and drawing upon their Marxist theoretical roots, "regional interaction is the exploitation of peripheral labour by core capital" (Cater and Jones, 1989, p. 20).

The preceding chapter had identified how various crises of profitability within the economic system (or more generically called 'crises of capital') had been the imperative behind shifts in production, including the advent of flexible labour contracts and the spatial relocation of investments to low wage or other similar low-cost states. Binding the capitalist system to a geographic system, Stutz and de Souza (1998, p. 545) describe how

> The buildup of crises in the capitalist core, frequently expressed as conflicts between capital and labor, has forced firms to emphasize multinational production and channel capital in the direction of peripheral regions, where hourly wage rates and fringe benefits are low.

The emphasis in the preceding paragraphs on the role of capital and transnational corporations is not accidental. World Systems Analysis illustrates the shift in relative power from states to capital. As Wallerstein argued, the critical organizing construct is not the nation state but rather the capitalist economy and its particular mode of production. As a result, society may be understood or defined "by the spatial extent of these modes of production, and not necessarily by political borders. Hence, ... the modern world system is a single society, the global capitalist world-economy" (Flint, 2009, pp. 813–814). In other words, there is an argument within World Systems Analysis that "another way to examine imperialistic relations is to focus on the role of the [multi-national corporation] as a source of political and economic control" (Stutz and de Souza, 1998. p. 545).

Dependency theorists also focus upon the relative inequality of bargaining power between multi-national corporations and host nations seeking to mobilize their natural resource wealth into economic and social benefit. As Stutz and de Souza (1998. p. 454) highlight,

> multinationals are imperfect organs of development in the developing countries, and their potential for the exploitation of poor countries is tremendous. There is, therefore, an inherent tension between the

multinational's desire to integrate its activities on a global basis and the host country's desire to integrate an affiliate with its national economy. Maximizing corporate profits does not necessarily maximize national economic objectives.

They add that while the relative bargaining power of host countries has increased over time, globalization also has increased the number of potential host countries competing for multi-national investments. In the 1980s, Japanese steel manufacturers were faced with high costs and limited supplies of high-quality coking coal. In response, they encouraged coal mine development in many countries including Canada and Australia. As this new production came on stream, the price of coal on the global commodity markets plunged and the steel manufacturers enjoyed the benefits of competing suppliers engaged in a 'race to the bottom' in pricing. Beginning in 2013, Asian power corporations were faced with transitioning away from nuclear (Japan and Korea) and older ('dirty') coal generating plants (China) and they looked to natural gas as a bridging fuel to a greener future energy system. Similar to the first story, they have encouraged many countries to enter the 'liquefied natural gas' (LNG) industry to the point where many huge projects were being planned around the Pacific basin. As the first of these came on stream, the price of LNG in the Pacific market began to drop and proponents in many countries were turning to the respective states for tax or regulatory relief to help keep their proposals competitive and potentially profitable.

Geographic models of spatial systems

The preceding section described how the global economy functions as an integrated system. This is not a new phenomenon, but the advent of modern communication and transportation technology has had the effect of 'shrinking' distances to the degree that the global economy is quantitatively and qualitatively more intertwined and connected than at any other time of history. The preceding section also highlighted that within the global capitalist economy the need for profit and the drive for capital accumulation necessarily means segmentation of activities and accumulation. This results in patterns of uneven development and places/regions that are dependent or subordinate within the system. As noted, resource-producing regions are dependent regions within the global economic system – dependent on the prices and demands set by the needs of the advanced manufacturing centres.

This section introduces a geographic model that has been useful for understanding such dependent relationships. The model was developed, and has been used, to describe the particular circumstances of remote, rural, and small-town resource-producing regions within developed economies. As we shall see, the model is 'descriptive' in nature in that it describes patterns and relationships but does not explain why they form or why they persist (that was the focus of the first part of this chapter). As we shall also see, the model is flexible in that it can be deployed at a range of scales from the local to the global.

'Core–periphery' and 'heartland–hinterland'

For our purposes, 'core–periphery' and 'heartland–hinterland' are not just word pairs linking opposites, they are the names of various descriptive models developed to help understand the relationship between metropolitan centres and resource-producing regions. Early work by Canadian historians such as Careless (1989) built an understanding of Canada's role as a raw resource supplier. As such, Careless sought to illustrate the enduring geographic relationships that characterize the nation's political economy. At its most basic, core–periphery models posit that political and economic control is maintained by a core region, which uses its historic structural advantage to organize and manage resource-production activities in a much larger geographic periphery. The wealth generated by the periphery thus supports the core and is directed for the benefit of the core. The flow of benefits in this way fits our earlier descriptions of the necessary prerequisites for capital accumulation in select locations and by select interests. It also fits with descriptions of the exploitive relationships that lead to uneven development as described with in World Systems Theory. As Markey et al. (2012, p. 43) note, "as long as economic and political decision-making power remains concentrated in the core, opportunities for diversification and advancement in the periphery will remain limited".

The key in such core–periphery / heartland–hinterland models is their focus upon the relationship between geographic regions within a wider social, economic, and political system. As described by McCann (1998, p. 8),

> In the heartland–hinterland system, regions interact with, or relate to, each other in a variety of important ways to shape their geographic character. The most important is usually economic, exemplified by exchanges from core to periphery of financial capital, goods and services, and government transfer payments, or by the shipment of staple commodities from a resource periphery to an industrial core.

But there are also wider impacts from this relationship, and these unfold and are reinforced over time: "Interaction between heartland and hinterland regions, whether of an economic, social, cultural, or political nature, therefore has considerable force in shaping regional and national character" (McCann, 1998, p. 9). Linking to earlier discussions about the imperatives of the capitalist economic system and the inherent uneven patterns of development that result, McCann (1998, p. 2) saw Canada's polarized heartland–hinterland geography as "a product of capitalism itself".

At its simplest, and following along from writers such as Careless and Friedmann, a metropolitan heartland holds control over a vast resource-producing hinterland. Heartland locations typically tend to be those which were the places of original settlement within regions which were chosen for a range of geographic advantages, including access to wide transportation networks and export opportunities through facilities such as ports. Heartlands are the connective piece between resource-producing regions and international markets. This is because they provide access to markets, including local markets as a result of their metropolitan size, and export markets because of their location at transportation and / or port hubs. Heartlands are also the managing partner in the global heartland–hinterland relationship as a result of their command-and-control functions that include government, legal, insurance, finance, and management. In other words, heartlands are the centres of political and economic power. As a result, they are able to maintain a structural advantage over hinterland regions (there are additional reasons for the lagging status of resource-producing hinterlands that will be discussed in the next chapter).

That heartland centres are geographically removed from the economic activities of hinterland regions does create problems for the system. This geographic separation can hide the state's economic dependence on natural resources (Baxter and Ramlo, 2002). This can create a false sense of security when it comes to interpreting global economic change. It can also act to isolate decision makers, and major media outlets, from the circumstances of hinterland communities and the role they play in the economic health of the entire system. This can be especially problematic with respect to investment decision making. If heartland residents and decision makers believe the urban core is the driver of the economy, they may invest heavily in its infrastructure without realizing the crippling long-run impacts of not re-investing in the social and physical infrastructure of the resource-producing communities and regions that actually do power the economy.

Hinterlands, as we have noted, are places of primary resource production. They tend to include small settlements that are usually located close to the resource in question. While heartlands and hinterlands are bound together in an economic relationship, growth and investment in the hinterland is heavily dependent on a continued resource extraction and basic processing. The processing is almost always limited to 'break the bulk' exercises that convert the raw material into a basic commodity. This might include reducing mineral-bearing rocks into ore concentrates or cutting trees into dimension lumber. With such a focus, hinterland regions remain economically undiversified and highly vulnerable to global commodity prices and market demands. There are limited opportunities for advancement in resource-producing regions so long as political and economic power rests in the heartland regions.

It is common to find clusters, or regions of similar type economies and communities, across hinterland regions. This is because the availability of the resources itself also tends to be concentrated in regions. Depending upon the resource, availability will be determined by patterns of climate, ecology, or geology. As noted, this results in hinterland landscapes that are comprised of numerous small settlements located close to where the resources are to be found. These settlements typically have small populations, are geared to supplying labour for local resource industries, and generally have only low-order goods and services available.

The hinterland's urban structure is typically a loosely organized system of small places. There are some interregional ties due to commonalities that arise from being engaged in similar economies, but most settlements have more direct and strong ties with the heartland (policy decision making by government and economic decision making by corporate head offices). Individually, they exercise relatively little economic or political power; instead, they are very much dependent on decisions made in heartland locations.

Closing

The previous chapter introduced the capitalist economy and suggested its functioning as a system. From that base, this chapter extended the introduction by describing the nature and functioning of a global capitalist system. As with the previous chapter, the fact that capitalism is a social and political, as much is an economic, process is important in understanding not just issues such as accelerated globalization, but also the implications of such for remote, rural, and small town resource-dependent places and regions within developed economies.

The chapter opened with an introduction to globalization. While not a new phenomenon, globalization has certainly been transformed and accelerated in the post-1980 era by the coincident processes of new information technologies, crises in Fordist-style capitalism, and the Neoliberal policy turn that fostered the discourse of support for international trade agreements. Linking our understanding of capital to the discussion of globalization, themes of dependence and exploitation highlight the resulting uneven geographies of investment and disinvestment.

The discussion of globalization then raised the political economy concern for the structures (political, social, economic, etc.) that support the global capitalist process. These include international trade agreements, but also international institutions such as the IMF, and the World Bank, together with international processes such as the Bretton Woods Agreement and the GATT process. For resource-dependent places within developed economies, the implications of globalization and the new international division of labour have focused on job losses, downsizing, and closure.

To help explain how these processes come together, we turned to World Systems Analysis. As a counter to earlier modernization theories that helped to backstop colonialism and more contemporary neo-colonialist intrusions into the political economy of developing states, World Systems Analysis married a Marxist analysis with the spatial framework of uneven development. Important for our understanding is the re-organization of dependency and domination. We then took these themes and further reviewed the processes they engender via dualistic spatial relationships of exploitation – specifically cores–peripheries and heartlands–hinterlands.

Taken together, the three sections of this chapter help advance our thinking about resource-dependent places and the changes they are experiencing. Globalization has opened up more opportunities for flexible and mobile capital to exploit cheap sources of raw material or labour, or the benefits of low regulatory regimes. Understanding the globalization of capital through World Systems Analysis allows for the identification of continued uneven development even where there might be significant industrial investments. Especially, it allows us to identify an explanation for dis-investments in older resource-dependent places in developed economies. These processes of dependence and exploitation are then reflected in the core–periphery / heartland–hinterland models of spatial relationships (which can be applied at numbers of scales).

To this point, we can see the patterns of uneven development and the imperatives of the capitalist system. Missing, however, is an explanatory framework for why resource-dependent places within developed economies were not able to transition to more advanced and diversified economic structures, even though they had experienced decades of significant industrial activity and its associated economic and employment spinoffs. Why are these places 'stuck' at the same level of development and why are they so vulnerable to the types of transitions that took place after the 1980s? It is to these questions that the next chapter turns.

5 Resource regions in a global capitalist economy

Introduction

Resource-dependent regions exist and function within the global capitalist economy. As such, we need to bring in frameworks that allow us to 'situate' those regions within the functioning of that economy and system. Most resource-dependent regions are organized around the export of a raw material or natural resource. In simple terms, the local production of low value-added resource commodities is exported to more advanced or developed manufacturing regions. Chapter 3 identified the nature of the capitalist economy and the imperative for profitability as the motivation for capital investment. In resource-producing regions, the opportunity to access a relatively low-cost natural resource, add incremental value, and then ship in large volumes creates just such an investment opportunity for capital. Chapter 4 then highlighted that the accumulation processes within the capitalist economic system both require and reproduce patterns of uneven development – with large peripheral regions being organized for the economic benefit of core regions (or within a World Systems Analysis for the benefit of transnational corporations).

Over time, however, these resource-dependent regions show many signs of being 'stuck' in the same level of the global economic system. While there are certainly exceptions, many others remain 'locked' in place until some catastrophic change (such as industry closure or resource exhaustion) forces a break from the past. Some places have worked through such breaks to create new economies, others have not. So how can we understand this 'stickiness' of the resource extraction pathway? To help with an answer, we turn to an older Staples theory literature and some more recent additions to the economic geography literature on 'evolutionary economic geography' and 'institutionalism'.

This chapter is comprised of two parts. In the first, we review the critical theoretical frameworks of Staples theory, evolutionary economic geography, and institutionalism. In the second part, we provide a number of case study examples of development and change in staples-producing economies on the periphery of the global economic system. The examples help to situate the colliding theoretical frameworks of a Marxist critique of the capitalist economy, the World Systems Analysis of global relations producing and perpetuating uneven development, and the challenges of dependency, truncated development, and path dependency illuminated through Staples theory, evolutionary economic geography, and institutionalism.

Staples theory

Harold Innis was the first to describe the political economy of Canada as being organized around the extraction and export of minimally processed raw materials to more advanced industrial economies (Tonts et al., 2013). Innis emphasized the concept of 'staples' to describe these raw material inputs and went on to illustrate how the Canadian economy had developed over time through a succession of resources that included fur, fish, gold, forest products, minerals, energy, etc. (Innis, 1956). As a supplier of raw material inputs, Canada's historic position relative to global trade has always "been one on the margin of the leading sites of global industrialism and innovation" (Haley, 2011, p. 99).

Innis also brought geographic space into the mix. His linking of economic process with geographic space also sought to include explanatory processes that underscored the dynamic core-periphery relations of staples-producing regions. As described by Argent (2017b, p. 20),

> For Innis, *geography* plays a central role in shaping the developmental fortunes of staples-dependent regions. A number of separate but related dimensions of space and spatiality are relevant here. First, Innis explicitly recognised the crucial role of the physical resource base (e.g. underlying geology) in dictating the location of the primary stages of resource extraction. Second, and relatedly, staples theory incorporates an understanding of the multifarious roles of the broader physical environment (e.g. climate, topography) in both enabling and constraining access to resources, labour, and higher order goods and services.

As Barnes (1996, p. 222) argues, Innis's geography "Is not simply a crude core–periphery model; its internal geographical differentiation is a necessary explanatory part of the wider model". The complexity of contemporary resource-dependent regions is the result of both historical processes and the changing forces of geography, technology, and institutions that shape and reshape the political economies of these regions. In terms of historical processes, Barnes (1996, p. 222) describes how there is a "layering of past and present rounds of staples accumulation".

The processes of change that impact the political economy of resource-dependent places and regions comes from the instabilities embodied within each of the noted forces of geography, technology, and institutional arrangements. As described by Barnes (1996, p. 222), "changes in geographical factors and their economic expression cause instability. Non-renewable resources run out, cheaper supplies are found elsewhere, and lower prices on international commodity markets make current production sites unfeasible". In terms of technological change, the adoption of new production methods or transportation strategies create disruptions and instability, while for institutional structures "The oligopolistic character of many staples firms necessary for the large capital expenditures produces financial and production 'rigidities' that cause disequilibrium" (Barnes, 1996, p. 222).

For Innis (1933), Staples theory helps to describe the social, political, and economic implications of natural resource dependence (Horsley, 2013). More specifically, Staples theory describes patterns of uneven development and the peripheral role states have assumed within the global economy (Sheppard, 2013). While other economic theories (i.e. modernization theories from Chapter 4) predicted that staples extraction would propel the economy towards maturation through the development of higher-order industries and interdependence, Innis countered that diversification and maturity were neither automatic nor assured (Marchak, 2011). Indeed, Innis's Staples theory has been labelled pessimistic because he saw the institutional framework associated with staples production as posing long-term barriers to development (Drache, 1991).

For Innis, a dependence upon staples reinforces a "centre-margin relationship [that] generates a series of self-reinforcing rigidities" (Haley, 2011, p. 100). The rigidities first set in place a socio-technological system to support the export of staples, a system that needs attention when crisis disrupts the staples market. As Haley (2011, p. 102) states:

> When crisis and change disrupt the old centre-margin relationship, the margin most eagerly seeks new connections with a centre to solve its internal problems. It must fill the unused capacity, pay off debts, and assuage its vested interests. It exploits more staples, often the different types of staples that the centres of the global economy demand, to enable new phases of growth ... [in other words] ...The staples trap describes the circular pattern behind Canada's historic tendency towards resource lock-in and truncated technological development.

Through Staples theory, the economic history and geographic unevenness of development in numbers of developed economies can be viewed as being shaped by the export of natural resources to meet the needs of national and international markets. Given the abundance of natural resources, it is not surprising that in countries such as Canada, Australia, New Zealand, and the Nordic states, national and provincial/state governments have long based economic development strategies on resource exports. Within this resource industry context, however, instability is entrenched (Ivanova, 2014; Nelsen et al., 2010). Two key problems that flow from a staples-based economy include 'dependency' and 'truncated development'.

The dependency problem comes from the resource supply 'warehouse' function of the economy. As 'price-takers', producing regions are always dependent upon the demands of external markets. As economically and industrially advanced regions expand or curtail industrial production, there are immediate impacts felt along the raw material supply chain. As Barnes (1996, p. 216) notes, "most staples regions tend to be vulnerable to demand shifts in markets that are both highly competitive and price-elastic". Barnes (1996, p. 216) further noted, this dependence upon volatile international markets produces "the characteristic boom-and-bust economy of resource-producing regions". Over time, the fluctuations in global commodity demand have become more dramatic – rising faster than previously experienced, going higher than previously experienced, and contracting at similarly fast speeds in today's hyper-connected global economy.

The dependence problem is exacerbated by the challenge of truncated development. World-scale resource-development projects require world-scale firms and access to requisite levels of financing. As we noted earlier, since the end of the Second World War, public policy in many jurisdictions has focused on attracting large industrial capital to realize the potential for massive resource industry projects (Fogarty and Sagerer, 2016; Marchak, 2011). Once in place, the region's resource commodity production path becomes increasingly entrenched as industrial capital consolidates its control over a stable and predictable supply of raw resources, its managing role on other components of the needed service and supply chain, and its dominance of local and regional labour by being the 'only game in town'.

As Hayter (1982, p. 281) argued, the implication of such foreign ownership "relates to a loss of autonomy over strategic investments and technology decisions" (see also Haley, 2011). Halseth and Sullivan (2002, p. 258) expand upon this idea, arguing that, "foreign controlled firms are often content to continue exporting basic resource commodities that are needed in their home economies or for other components of their multi-national holdings". While both of the preceding quotes refer to foreign owned companies, they might be better written today as referring to transnational corporations. Regardless of where the company is 'headquartered', but rather that they are transnational corporations with global holdings and responsibility not to places or regions but to shareholders.

One key effect of truncated development is that the long-held mantra in resource-producing regions about the need to diversify within existing sectors, across existing sectors, and to add whole new sectors, is nullified by the dominant industry's requirement to remain the regionally dominant industry. Long-term shifts to slowly transform a dependent region into a diversified region are trumped by 'crises' within the dominant industry that threaten massive layoffs of labour and loss of tax and royalty revenues unless assistance and concessions are granted (Wilson, 2004). The 'too big to fail' ethos has applied for decades in the industrial resource-development version of staples economies (Marchak, 2011).

In a 'back to the future' moment and policy initiative, many governments recently were part of the 'rush' to launch aggressive efforts to attracting an

LNG industry to their jurisdiction. With declining market demand in the US and increasing market demand in Asia, the BC government followed a very Neoliberal policy idea about 'freeing' BC's supply of natural gas by bringing it to ports along the Pacific coast so that it can be shipped anywhere in the world. However, as described in Chapter 4, the decline in global prices and the fact that many natural gas producing states are also trying to get into this market means that proposed industry developers in BC (transnationals such as Shell and Chevron, and national firms such as Malaysia's Petronas) are calling for concessions in order to create 'competitive' environments and this includes reductions in taxes and costs. While benefits to the state, the region, and to local communities are being eroded even before the first plant is built, the development pathway will be 'set' for a long period of time due to the commitment of the resource to a single industry – historic patterns of dependence and truncated development will be replicated. Plummer and Tonts (2013) and Argent (2013) describe similar 'booms' in resource production across different regions of Australia that form the core of a re-emergence of staples driven national and state development policy. As Tonts et al. (2013, p. 365) identify, "public expenditure in resource regions also remains comparatively low, and indeed tend to reproduce a development framework oriented towards extraction rather than diversification".

Fordism to Flexible production

While not unique to industrial resource commodity production, the impacts of the shift from a Fordist to a more Flexible mode of production are important, and extend the staples discourse into issues associated with the local and regional benefits that might be derived from resource exploitation. As noted earlier, resource commodity production has since the 1980s been undergoing a transformation, or restructuring, similar to that found in other production sectors. Sometimes cast as a response to a crisis of capitalism or to the rise of competition from low-cost producer regions through increasingly globalized markets and international agreements that reduced protectionist barriers to trade, the general result of the transformation is significant employment reductions and ongoing negotiation by capital to lower the regulatory costs of production.

Most post-Second World War industrial resource commodity projects in developed economies were seen as a significant tool for both regional and national development (Bayari, 2016; Mitchell, 1983; Nel, 2015). The massive scale of production provided equally massive revenue flows to central governments through a host of tax, fee, and royalty schemes. They also provided for large numbers of jobs which stimulated and supported local economies where the resources were being extracted (something which further increased revenue flows to central governments through various income and consumption taxes). Through local and regional procurement of goods and services, these pre-1980 industrial resource developments also supported growth in regional economies.

Following the global economic recession of the early 1980s, the historic pattern of employment and benefits flows has changed radically. The older Fordist mode of production supported large workforces distributed across highly differentiated tasks in a repetitive, mass production style, approach to resource extraction (see Chapter 3). Since the early 1980s, however, these industries have followed the trend towards a more Flexible mode of production. The reorganization of work towards multi-tasking, and the wholesale replacement of labour with increasingly advanced and computer controlled machinery and technology, has meant job losses in resource-producing regions that had previously experienced decades of employment growth and stability (Halonen, 2017). In resource-producing regions, it is now common that less employment and fewer local benefits are flowing from raw materials processing even though the volume of exports is increasing.

In his essay for the *Dictionary of Human Geography*, Barnes (2009, p. 722) explores how Staples theory works to account for ongoing economic dependency and instability. Root causes for these tendencies include the previously mentioned points that staples regions are price-takers in the marketplace, and that innovation and import substitution put long-term downward pressure on commodity demands. Barnes also describes a number of issues arising from the dominance of large transnational corporations in resource production. Among these issues are that states undervalue resources as part of efforts to attract and hold economic investment, that without value-added processing the benefits derived from this already discounted resource are small, that there is a loss of control over resource production, as well as the previously noted point that such firms have no incentive to diversify or add value to a raw input that will be used elsewhere in their global production or marketing chain.

Drawing on a political economy framework that links social and economic development to the type and form of staples production, Barnes (2009, p. 722) also notes how Innis "brought together three types of concerns: geographical/ecological, institutional and technological." Particular sets of geographies, institutions, and technologies collide at moments in time to create intense levels of activity.

Evolutionary Economic Geography

With some focus upon the construct of 'regional path dependence', the literature and debate within evolutionary economic geography can offer a different yet aligned framework to Staples theory. Most especially, recent discussion around Martin's (2010) call for a "rethinking of path dependency in economic geography that underscores the importance of change" (Simandan, 2012, p. 156) provides a counterpoint to staples dependence and a set of tools for evaluating whether transformation in regional economies is one of emergence and transition, or simply one of continuity.

Argent describes evolutionary economic geography (EEG) as being different from neoclassical economics in its treatment of both space and time. He writes how:

> in contrast to [neo-classical economics'] treatment of space and time as passive dimensions within the economic development process, EEG's ontology regards these two fundamental dimensions as intertwining and interrelated active forces shaping societies and economies. Whereas orthodox equilibrium economic models downplay the significance of historical events and processes of spatial competition in moulding the economic landscape, EEG research accentuates the roles of technological innovation, bounded rationality among economic agents (e.g. entrepreneurs), and the role of institutions in driving economic change.
>
> (Argent, 2017b, p. 24)

As with Staples theory, the particular alignments in time of space / place, institutional structures, and technological advances loom large in explaining outcomes.

For Boschma and Frenken (2006, p. 277), "Evolutionary Economic Geography applies core concepts and methodologies from evolutionary economics in the context of economic geography." Starting with the view that systems are sets of embedded routines; routines in firms, routines in policy, and routines of practice in institutions, EEG then turns to the issues of change and innovation at local and especially at regional scales (Suorsa, 2014). As befits its connections to geography, there is a concern with the spatial distribution of routines, and changes in those routines. As Martin (2012, p. 180) has argued, "any convincing theory of regional development needs to give explicit recognition to, and account for, the roles of history and path dependent dynamics and outcomes." For Argent (2017b, p. 24),

> While path dependence helps to explain how a branch of industry becomes 'locked-in' to the use of a specific innovation, for economic geographers *path* dependence is often equated with *place* dependence in which local or regional economic processes can be said to be, in one way or another, substantially dependent on or correlated with past events and trajectories.

But rather than viewing path dependence as inevitably supporting 'locked-in' regional economies, Martin (2012, p. 183) argues that there must be room to admit other "more incremental and developmental patterns of evolution". This mix of stability and change is recognized by Fløysand and Jakobsen (2016) who identify that EEG studies have emphasized both continuation and change.

But how do changes occur within path dependence? Are they slowly evolving, are they sudden, or are their sources internal or external? These questions are the subject of much debate within EEG (Oosterlynck, 2012). Writing from an economic perspective, Bowles (2013) has noted how the northern BC resource commodity economy had adjusted through at least three phases – each with quite distinct policies and consequences. He notes that at present, northern BC is "experiencing new globalizing patterns as trade and investment flows and economic structures adapt ... [to the rise of the Asian economies, and] ... is actively seeking to reposition itself within the multi-centred global economy" (Bowles, 2013, p. 261).

However, EEG, and its antecedents in new economic geography and even neoclassical economics, imports expectations that agglomeration economies will create positive 'spinoff' effects. These include inter-firm competition, support and supply chains, labour arrangements, knowledge and technical production, etc. As with Staples theory, social-political-technological arrangements become established and then become embedded within temporally bounded regional economies. Also as with Staples theory, change over time is both expected, and expected to be circumscribed by past economies and dependencies. Using EEG ideas to explore processes of change in rural Australia, Tonts et al. (2012) found that evolutionary concepts like place and path dependence help us to understand the deep historical and spatial processes that lead to the creation of particular rural places as bundles of 'space / time'. This attention to context fits, for example, with our broader interest in how place-based approaches to regional development can assist in understanding the nature of transformation and change over time (Markey et al., 2012).

The Pilbara region in western Australia is one that has received a lot of attention *vis-à-vis* the application of an EEG approach to understanding. In the introduction to his study of the Pilbara, Peck (2013, p. 243) describes the area as:

> a vast and sparsely populated desert region, the Pilbara is host to some of the most profitable and capital-intensive extractive activities on earth (based on the exploitation of iron ore, oil and gas, and manganese), and epicenter of the global resources boom. Known as 'China's quarry', the region relies on a large and richly remunerated workforce of 'fly-in/fly-out' (FIFO) contractors, its transitory boomtowns awkwardly coexisting with Indigenous communities hovering on the brink of social crisis.

In a general sense, the Pilbara region has experienced the opportunities and challenges common to resource-dependent rural and small town places and regions. As noted by MacKinnon (2013, p. 317), resource-led development in the region has "resulted in development 'in' rather than 'of' the region with many of the benefits being gained by external interests such as capital, labour and the WA state rather than most groups within the region". Within that general context, the research literature has noted how the contingencies of place continue to matter. In the interplay between global capital and local interests, MacKinnon (2013) highlights the 'unbalanced' or unequal nature of the relationship. Tonts et al. (2012, p. 288) highlight how the distribution of socio-economic benefits and well-being is "highly variable, contingent on a range of factors including the nature of the local commodity, company structure, and location". While evolutionary aspects of development transition are evident, they are also not equally distributed across the region. Plummer and Tonts (2013, p. 239) conclude from their studies that the results begin:

> to tell us what geographers know intuitively: local context matters. Each of the sub-regions examined here is following a quite distinctive evolutional trajectory with respect to income and employment. This is despite the fact that they are all being embedded within the same regional 'resource conflict'. We would argue that these evolutionary pathways reflect, *interalia*, the different natural resources involved in each location, different company structures, and institutional arrangements in place locally and governing the resource ... and particular social and cultural dynamics.

In EEG, clearly place and contingency still matter tremendously.

Institutionalism and economic geography

The third theoretical framework we introduce in this chapter to support our overview of perspectives useful for understanding change in the remote, rural, and small town resource-dependent regions of developed economies is 'institutionalism'. Drawing upon Hayter (2008), the institutional perspective has a focus on 'techno-economic' paradigms. With a long theoretical pedigree that reaches back to the 19th century and the work of Thorstein Veblen, institutionalism has emerged as a counterpoint to a Marxist analysis of the capitalist economic system. Instead of following the generalized, and perhaps totalizing, imperatives of capital and the labour-capital dynamic, evolutionary institutionalism:

> rejects the idea of a universal model of capitalism and describes its generalizations and prescriptions from close observation, measurement and interpretation of real-world institutions that are variously expressed as formal and informal organizations, movements, common values ... an insistent theme for evolutionary institutionalism is the power of big business to shape the nature of market economies, rather than simply passively responding to consumer sovereignty and government regulations.
> (Hayter, 2008, p. 834)

For our purposes, the institutional approach reinforces the need to identify the international capital players, in the form of transnational firms, in the economic restructuring of resource economies.

Institutional economics is described as a 'third way' or middle approach to understanding and characterising the economy (Hudson, 2006). Prior to the economic restructuring of the 1980s, economic geographers commonly approached the economy using either neoclassical or Marxist theory (Cumbers et al., 2003). The neoclassical approach assumed that economic activity is rational, maximizing, and atomistic, while ignoring complex arrangements of social and political forces, space, and people at work (Martin, 2000). Marxist approaches emphasized the importance of social structures while placing little emphasis on individual and collective agency, or the role of culture (Cumbers et al., 2003). Both approaches are under-contextualized in that the peculiarities of places and society are not addressed (Martin, 1994). The institutional approach has gained some favour because it accounts for the social and cultural conditions of everyday life and real world behaviours (Hayter, 2004). Institutionalists have focused on how past and present arrangements of individuals, institutions, cultures, and politics in different places work to shape economic development (Peet, 2007).

Institutions are socially constructed and include formal regulations and organizations and the informal practices, routines, and habits that shape the behaviour of actors. Formal regulation includes laws, policies, standards, and rules. Informal or tacit practices can include local networks, habits, norms, routine patterns of behaviour, and customs (Streek and Thelen, 2005). As such, the make-up and mix of informal and formal institutions are rooted within place, and tend to vary across space (Boschma and Martin, 2010). The strength of this approach for studying regional economies is that it takes a broad view of economic activity, where different actors (firms, governments, organizations, individuals, etc.), different imperatives (within the capitalist economic system), and different operating contexts (within a global economic system), cannot be divorced from local culture, history, or society (MacLeod, 2001).

Through this theoretical lens, regions take on a new prominence as meeting places for a variety of global and local institutions and the conflicts, relations, and discourses that shape, constrain, or transform the economy, society, and culture (Paasi, 2009). This conceptualization is grounded firmly in the notion that regions are not fixed, concrete, or predetermined entities (Lee, 2002). Rather, regional economies are perpetually fluid due to the changing nature of resources, markets, technology, policy, and relationships over time (Essletzbichler and Rigby, 2007). This is an important distinction because it raises the role of space, scale, and time in shaping the particularities of different regional economies (Page, 1996). Variation across regions is expected, and the development of regional economies will follow a trajectory of change shaped, in part, by the history of past decisions (Peck, 2005). The new institutions are built, or layered, upon existing institutional arrangements.

The 'evolutionary' aspect of evolutionary institutionalism links to the concept of change (like with evolutionary economic geography). It makes this link through the analysis of innovation. For Hayter (2008, p. 831), evolutionary institutionalism "interprets industrial transitions through the lens of innovative behavior that is shaped by reciprocal economic and non-economic processes and periodically restructures economies in the form of new techno-economic paradigms (TEP)". The concept of innovation provides a "central, unifying theme that co-evolves as institutional and technological dimensions according to changes that are incremental, major, radical and paradigmatic" (Hayter 2008, p. 832).

Clapp et al. (2016) examine one dramatic experiment with institutional innovation in their study of the Great Bear Rainforest agreement covering lands along the mid-coast of British Columbia. As part of a transition from government to governance over natural resource development, the agreement includes many interests, including the federal and provincial governments, Aboriginal peoples, environmental interests, industry, and local non-Aboriginal communities. The complexity of the stakeholder dialogue is reinforced through at least six bi-lateral sets of negotiations between the different signatories. While the products of these negotiations have precedent elsewhere, Clapp et al. (2016, p. 255) suggest that:

> they collectively represent institutional thickening in many dimensions on an unprecedented scale in the thinly populated region ... The outcome of this remapping process is neither a free market solution, nor public control, but an increasingly complex architecture of institutions, based in both civil society and the state, that promote sustainability, resilience and legitimacy.

Whether such institutional thickening supports or restricts the flexibility that rural regions need to respond within the fast-paced global economy remains to be seen.

Innovation and evolution also link to the EEG framework. As Boschma and Frenken (2006, p. 289) point out, within "an evolutionary framework, the key issue is to analyze the extent to which institutions are flexible and responsive to changes in different places ... [and how] ... institutions co-evolve with processes of technological innovation and industrial dynamics." This role for innovation also connects with our concern with assessing the economic transformation of resource regions in terms of transition or continuity. Just as Staples theory understands that resource commodity economies will emerge from, and will co-create, unique combinations of technology, policy, and social constructs to support the particular mode of production, so too does the evolutionary institutionalism approach recognize the temporal and geographic collisions that create unique TEPs.

Staples economy examples

In this second part of the chapter, we include a number of brief examples of staples economies. As noted, the policy framework in many OECD states during the immediate post-war period supported industrial resource expansion into remote and rural regions. As the challenges of the global economic recession of the early 1980s and the subsequent pressures around restructuring began to unfold, the policy responses of OECD states was also quite similar – retrenchment of investments in resource-dependent regions and policy change to support industry competitiveness and thus ensure a flow of revenue to the state. The purpose of each of the examples below is to set the stage for understanding the impacts of restructuring. While there are changes, there is also a good deal of persistence in terms of adhering to a staples approach in each of the peripheral regions profiled.

A Finnish illustration of a staples economy

Change in rural Finland has been profound for the industrial, employment, and community landscapes of staples-producing regions (Kotilainen et al., 2017). These have largely been driven by corporate restructuring strategies, changes in regional development policies, the impacts of global economic recessions, and changing relationships with Russia and the European Union. Many peripheral rural regions started to industrialize in the late 19th century with small forestry holdings leading to the development of forestry processing. Depending upon the region, there were other nascent industrial activities. A further expansion of the forest sector through manufacturing and engineering emerged in the 20th century, in part due to its proximity to European and Russian markets and its territorial advantage of abundant raw materials (Tykkyläinen et al., 2017). This growth was spurred by many decades of national ownership and policies that were strategically deployed to promote the development of remote rural regions.

After unsuccessful attempts to expand globally in the mid-1960s, many Finnish forestry companies expanded internationally around the mid-1990s and shifted production with investments in the United States, Asia, and South America where lower labour and material costs could enhance their competitiveness (Tykkyläinen et al., 2017). There was also growth in company mergers, including the merger of Enso-Gutzeit Oyj with Stora AB. The collapse of the Soviet Union in the 1990s further meant that new markets for Finnish forest products had to be developed. As a recession took hold in the early 1990s and foreign debt rose, the national government lost some of its capacity to "influence industrial price competitiveness by devaluating the currency after wage inflation" after joining the Economic and Monetary Union and introducing the Euro (Halonen et al., 2017, p. 202). At the same time, debates about forest biodiversity and forest management practices meant that environmental issues continued to shape the forest sector in Finland. In the 1970s, for example, concerns about environmental impacts and health concerns led to strategic investments in new production technology; spurred by the Finnish government's regulations for lower emissions and closed processes (Tykkyläinen et al., 2017). By the 1990s, an industry-led Programme for the

Endorsement of Forest Certification (PEFC) was favoured over other third-party certification standards.

Throughout the early 2000s, Finnish companies were looking abroad to invest, while foreign investors were purchasing industrial assets in Finland's periphery. By the latter half of the 2000s, industry was increasingly out-sourcing maintenance and production tasks (Halonen et al., 2017). The combined result was extensive layoffs and closures of mills in Finland's rural regions. It was during this period that the Finnish government also sold off shares in state owned companies.

The Finnish government has pursued strategic planning policies since the mid-1970s (Kotilainen et al., 2017). These policies were reshaped and reformed by Finland's entrance to the European Union in 1995 that connected the country to EU structural adjustment funds such as those in the LEADER policy programme. Through these new geopolitical arrangements, much of the responsibility for regional development was transferred to resource-based regions under the auspices of bottom-up local development that sought to encourage rural actors to think at a European scale. This Neoliberal shift in responsibility was executed quite differently than in other states because of the extensive Keynesian backdrop of top-down EU policies and funding support.

National growth policies in the mid-1990s were unsuccessful in addressing deficiencies and uneven development in Finland's peripheral rural resource-based regions. In fact, researchers suggest that many of these regional growth policies have been strategically targeted to nurture the development of a value-added service and information technology sector in regional centres (Tykkyläinen et al., 2017). The national government also introduced sub-stantial reductions in public spending, prompting services from post offices to border stations to be closed and regionalized. As in other states, such public sector efforts at budget / spending reductions negatively impacted rural regions much more than they did larger urban centres.

During the last decade, researchers have observed structural problems with Finland's mature forest industry; the result has been a decline in the share of the nation's GDP that the industry contributes (Tykkyläinen et al., 2017). Communities have struggled to stave off population and economic stagnation and achieve economic diversification. These efforts have been limited by a lack of skilled labour, low general levels of education, population aging and youth out-migration, and a loss of regional policy funding; all working to reinforce the path dependence trajectory of these places. Developments in small business and services have not been sufficient to curtail the loss of employment in other sectors. As they move forward, community stakeholders are looking to tourism, the second-home industry, and the biotechnology / bioeconomy sector to sustain their economies. For example, the Lieksa Development Agency is pursuing the development of a bioenergy plant near the Kevätniemi sawmill – pending licenses and financing (Kotilainen et al., 2017). There are concerns, however, that these resource-based regions lack access to appropriate skilled labour locally to pursue these endeavours, leaving educational institutions with the formidable challenge of quickly developing supportive training and development programs in order to better position these resource-based regions for the future.

An Australian illustration of a staples economy

Argent (2017a) has explored the transformation of Australia's agricultural sector through economic, labour, and community restructuring processes. Writing about New South Wales, he describes this transition as being from an era of productivism following the Second World War to one of multi-functionality since the 1980s. In the immediate post-war period, the Australian government implemented a series of policies and programs to expand and intensify agriculture and agricultural manufacturing. Much of this policy focus was designed to address concerns over international food shortages. This included the introduction of high levels of industry protection through tariffs and quotas designed to allow the national manufacturing sector to firmly establish itself as a key employer. Not all regions were able to develop a more diverse agriculture-based industrial and manufacturing infrastructure; and for these regions this reinforced economic vulnerability and path dependency.

The first signs of trouble for the agricultural sector were marked with the collapse of the international wool market in the 1970s. As the Labor government rose to federal power in 1972, there were cuts to tariffs, quotas, and other mechanisms of protection for resource sectors, including agriculture. This left agricultural producers, and their communities, increasingly exposed to market fluctuations. The national government's pursuit of Neoliberal policies supporting open and free trade markets continued throughout the 1980s, 1990s, and 2000s, as it continued to dismantle and remove institutions, mechanisms, and programs (e.g. Egg Marketing Board and the Australian Wool Corporation / wool reserve price scheme). As Argent (2017b, p. 33) writes, "concessions to farm productions (e.g. fertilizer bounties, subsidies for farm fencing, free agricultural research and development advice) and in support of rural settlement (e.g. subsidized water and electricity, maintenance of rural roads and railway lines) were wound back." Capital intensification and the adoption of labour shedding technologies designed to enhance global competitiveness were also having an impact on the agricultural sector and related employment.

As broad policy and structural changes were transforming the agricultural sector, Labor governments in the 1980s and early 1990s also began to initiate a series of Neoliberal policy approaches that would favour employers – starting with various accords concerning wage conditions that were tied to the need for increased worker productivity (Argent, 2017a). The transformation of the labouring landscape would be intensified through the Conservative-led coalition government (1996–2007) that reformed and largely removed the influence of unions while pursuing greater flexibility in bargaining arrangements. Since then, union representation and membership has continued to decline.

Despite a decline in farm employment, the out-migration of youth and inability to compete with industry wages in other sectors has impeded the development of the next generation agricultural workforce. In the 1990s, the agricultural sector's labour shortages led the introduction of new immigration and temporary visa programs (e.g. Temporary Business (Long Stay) Visa category 457 visa, the Working Holiday Maker visa program, and the Pacific Seasonal Workers Pilot Scheme). The Pacific Seasonal Workers Scheme, for example, has been effectively used to address farm labour shortages in the smaller agricultural towns of New South Wales (NSW). Temporary visa programs, however, have prompted concerns that they are being used "as a 'cheap fix' for short-term structural mismatches in the labour market, and to undercut Australian award pay and conditions" (Argent, 2017a, p. 151).

Despite rapid growth in mining activity, other resource-dependent regions have continued to decline as a result of industry's widespread use of fly-in / fly-out and drive-in / drive-out operations; impacting local and regional opportunities in wholesale, retail, accommodation, and food businesses (Argent, 2017b). The out-migration of youth and the aging population have also weakened the regional workforce and further challenged the capacity of community responses to new industrial and work environments. The capacity of small communities has also been challenged by their limited knowledge and ability to change and pursue new innovations that can better position their assets and economies (Sorensen, 2017).

Governments have attempted to address this issue by directing subsidies into research and development in order to nurture innovation and competitiveness across the agriculture and manufacturing sectors. Rural areas around Armidale, for example, have benefited from their close proximity to the University of New England (UNE) that has focused on animal genetics, innovations in agricultural products, and developing supportive research institutes. This regional centre also became the first community to complete the installation of the national broadband network. This eventually led Northern Inland to develop the Digital Economy Taskforce and the Digital Economy Implementation Group (DEIG) to help re-orient rural businesses to pursue an online presence. An important initiative under the DEIG was to persuade local information and communication technology-related businesses to switch from the Internet Protocol 4 to IP 6, making local businesses more compatible to enter trade partnerships around Asia (Sorensen, 2017). Looking forward, financial resources from the NSW government have been secured for a data centre that would function as a cloud-based information storage for regional businesses. UNE is also working to become a hub for GPS data to support agricultural and environmental applications. Early successes have included new products to detect environmental disasters, such as forest fires and floods, as well as new equipment manufactured in Armidale to assess soil moisture and nutrients, animal health, and quality of crops.

A New Zealand illustration of a staples economy

Focusing on the West Coast and Southland regions of New Zealand, Connelly and Nel (2017a, 2017b, 2017c) clearly demonstrate the complexity and impacts of industrial and employment restructuring on rural communities and regions. Since the Great Depression and through the immediate post-war era, the state pursued a Keynesian public policy framework with strong levels of management and intervention designed to address uneven development and regional disparities. There were high levels of state ownership of resource industries and these industries enjoyed strong access to British markets. These circumstances changed dramatically with the UK's entrance into the European Economic Community. In response to this loss of protected market access, the New Zealand state government adopted a suite of dramatic and sweeping economic and labour reforms (Connelly and Nel, 2017c).

Under a Neoliberal policy approach, pursued through the Labour government's 'Rogernomics', the state government removed many of the key policies that had supported the agricultural sector. The election of a more conservative national government simply continued the adoption of policy reforms throughout the 1990s. These included the privatization of state-owned companies and marketing control boards, an elimination of industrial / agricultural subsidies, the removal of collective bargaining rights and processes, as well as a reduction and regionalization of public services and programs. Two poignant observations by Connelly and Nel (2017c, p. 120) clearly demonstrate these transformations in public policy:

> Between 1973 and 2006, the primary sector's contribution to the GDP fell dramatically from 26 percent to 7 percent of the GDP, and non-food manufacturing fell from 19 percent to 11 percent as import tariffs were phased out, leading to the closure of 20 percent of manufacturing plants … in the 1990s.
>
> Between 1987 and 1990, the state-owned coal corporation reduced its employment from 1,861 to 715 jobs; the state telecoms company cut its employment from 24,000 to 16,263, Electricorp (the state-owned electricity supply company) went from 5,999 to 3,690 employees, the railroads from 14,900 to 8,400 and NZ Post from 12,000 to 850 … the civil service was trimmed from 62,102 jobs to 34,505 between 1983 and 1994 … in terms of postal services, 625 district post offices were closed, weakening rural service provision and employment.

The Business Roundtable, representing corporate interests, was successful in transforming the labour policy landscape, starting with reforms to the Labour Relations Act in the mid-1980s which moved away from nationally binding agreements to single employer agreements, therefore weakening labour's bargaining power and requiring all unions with fewer than 1,000 members to merge (Connelly and Nel, 2017b). While Social Support Coordinating Committees were established to assist workers and communities, the state government did not support these for very long and they were eventually eliminated in 1987 (Connelly and Nel, 2017a). By 1991, the Employment Contracts Act was introduced, further reducing the regulation of arbitration and weakening the right to strike. The weakened position of unions, however, also weakened the voice of small communities in this context of rapid change.

These changes in policy, and their impacts in rural and small town resource-dependent regions, were exacerbated by volatile fluctuations in commodity prices and the adoption of labour-shedding technologies that increasingly concentrated populations in regional manufacturing centres and reduced broader forestry, mining, and agricultural employment. The subsequent decline in farms and sheep herds also resulted in an overcapacity of slaughterhouses, prompting the consolidation and closure of numbers of plants. In fact, from the mid-1980s to the mid-1990s, the processing capacity fell by 25% and employment fell by 40% with significant repercussions for rural places (Connelly and Nel, 2017b). In the late 1990s, the forest sector also faced challenges such as when 84% of the West Coast region was converted from productive forest to protected forest for conservation purposes; leaving potential economic development options in dairy production and tourism. In response, the state government provided NZ$120 million to assist West Coast communities in transitioning. From this, NZ$92 million was invested into a trust called Development West Coast; however, many 'smokestack' investments and pursuits have done little to diversify the region's economy (Connelly and Nel, 2017a).

While the state has had little success in changing the staples-dependent nature of the rural economy and exports, some small towns have pursued tourism as a post-industrial development strategy. As well, improved dairy prices in the 2000s did support an expansion of the dairy industry in regions such as Southland. Agricultural producers are also diversifying their products through higher-quality meats and niche products in order to expand their market share and to find new markets. As in the Australia case above, the renewal of the dairy industry has meant that farms had to draw upon seasonal migrant workers, largely from the Philippines, to meet labour needs after local labour forces were degraded through out-migration and related restructuring processes. Public policy also facilitated a return of collective bargaining through the Employment Relation Act in 2000, but it did not support or strengthen any movement for unions, but rather focused on mediation (Connelly and Nel, 2017b). Union positions have also been weakened by the 2008 global economic crisis and the rapid growth of new mobile labour arrangements and the trend towards contract employment.

The state government attempted to roll out policies that would support business development centres and partnership programs to nurture collaborative arrangements; but these were largely unsuccessful. The trans-formation of New Zealand's resource-based economies has also been chal-lenged by their remoteness from distant economic centres, and an absence of government support to develop long-term regional development strategies. In the absence of supportive state policies, local governments have been left on their own to pursue investment and to ensure that appropriate services and supports are in place. Unfortunately, many local governments that urgently need to pursue new strategic directions for their economies remain positioned to manage existing assets rather than pursue a new entrepreneurial approach to transformation. As a result, small towns have remained locked into long-term structural decline. This has accentuated uneven development, with more urbanized centres benefiting from opportunities afforded through their ser-vice-based economies. As Connelly and Nel (2017a, p. 321) write, "state sup-port for economically struggling rural communities in New Zealand is virtually nonexistent, with the state focusing instead on growing regions and high growth potential business sectors, reflecting New Zealand's now long-term acceptance of its neoliberal policy."

An Icelandic illustration of a staples economy

The previous examples of transformation in the staples economies of resource-dependent regions have focused on long established forestry, agri-cultural, or mining regions and industries. But staples production is about mobilizing an abundant natural resource for export into the advanced manu-facturing processes of more developed economies. It is about transformation in the rural region of that resource into a form that is more amenable to easy transport. In this last example we look to Iceland – a country that is an energy 'superpower' but whose energy resources (principally hydro and geo-thermal) are not easy to export. As a result, Iceland has executed a policy to attract industries that require large amounts of energy that then convert the energy into a basic product that is easy to transport to off-shore up-stream manufacturing. In particular, we look to the example of aluminum smelting.

Iceland's first experience with aluminum smelters began with negotiations with Alusuisse to establish a production facility at Straumsvík in Reykjanes in the 1960s. Iceland's national government's pursuits were driven by a goal to diversify its economy that had largely been based on fishing. To accomplish this goal, Iceland promoted its advantage with a key natural resource asset – an abundance of hydropower. Iceland also offered companies political stabi-lity and an ideal location – within close proximity to both European and North American markets. Later, the country could also promote its European Economic Area (EEA) membership that would provide tariff-free access to the European market (Hilmarsson, 2003). Despite their natural advantages, negotiations were influenced by rapidly expanding aluminum operations around the world. Alusuisse, for example, had been expanding its operations in Africa and Australia (Skúlason and Hayter, 1998). There were also concerns about the rise of, and competition from, nuclear power that could impact negotiations. In 1961, the Icelandic government created The Industrial Development Committee to assess the potential viability of the aluminum industry and to lead negotiations. The potential cost of developing the power, however, meant that power supply and rates were key factors in the negotiations. Alusuisse would not provide any support to develop hydropower infrastructure. To finance such investments, the Icelandic government required support from the World Bank. Approval of World Bank loans centred on the economic viability of hydro rates of no lower than 2.5 mill per KWh, and the ability to obtain 33% of net profits from smelter operations through taxes (Skúlason and Hayter, 1998); thereby, defining Iceland's bargaining position.

By 1966, with considerable hydropower resources, the government offered Alusuisse low electricity rates for 15 years to encourage the development of an aluminum smelter at Straumsvík. Power rates were set at 3.0 mill per KWh for the first six years and reduced to 2.5 mill per KWh for the rest of the contract (Skúlason and Hayter, 1998). In terms of taxation, Alusuisse agreed to pay US$12.5 per tonne for the first six years and US$20 per tonne for the rest of the contract. The company also agreed to expand its smelter from 30,000 tonnes to 60,000 in production capacity in six years. In return, the national government agreed to waive import duties on machinery and materials required for the plant.

Location also became a key factor in negotiations. Despite a number of potential sites that could strengthen responses to its regional development policy and nurture growth away from Rekjanes, Straumsvík always remained a key site. It was able to offer lower construction costs. While the Icelandic government imposed stricter environmental requirements for northern sites (i.e. Dysnes), prevailing winds and offshore ocean currents dispersed pollutants, meaning that stricter environmental regulations did not have to be in place. Industry also benefited from close proximity to an international airport in Reykjavík to support access to broader local and international labour. National government support was also provided to develop harbour facilities in the municipality. There were also questions about supply of labour available in smaller, northern communities to support the construction of such a large project.

The government attempted to pursue higher rates after the initial agreement with Alusuisse expired in the early 1980s, only to be confronted with industry resistance and threats of closure (Kirchner, 1988). Alusuisse renegotiated to link power rates to aluminum commodity prices. By the 1980s, Iceland offered the lowest electricity rates in the world for aluminum smelting (Skúlason and Hayter, 1998). In Iceland, "the new smelters that were constructed reflected a perception that cheap electricity would enable a national aluminum industry to be internationally competitive" (Kirchner, 1988, p. 72). This was particularly important since the raw materials had to be imported.

Since the 1980s, Iceland established a committee to assess other potential locations to support aluminum production. By the early 1990s, the Icelandic government used its experience to negotiate the development of a smaller aluminum smelter facility operated by an American company, Columbia Aluminum just north of Rekjavik, in Halvfjöröur (Hilmarsson, 2003). There was also an expansion of the Reykjavik smelter, now owned by Alcan, and a second smelter built by Columbia Ventures on the west coast (Hilmarsson 2003). This was followed up with a new memorandum of understanding between the Icelandic government, Landsvirkjun, the national power company, and Alcoa to develop a new hydro power project in eastern Iceland that would support a new aluminum plant at Reyðarfjörður (Hreinsson, 2007). In their cost-sharing agreement, Alcoa agreed to share the financial risk incurred by the national power company for the power plant, in order to ensure that power would be available by 2007. The plant was in full production by 2008. By the 2000s, the national government engaged in more purposeful and formal arrangements through an investment agreement, a power contract, a site agreement, and a harbour agreement (Government of Iceland, 2002). The company was also required to complete an environmental model of the plant's emissions, an environmental impact assessment, and to obtain an environmental operating license for the plant; although it is not clear what modifications may have been made through these processes. The 40-year power contract, however, would provide 4,700 GWh/year, with a price review after 20 years. Alcoa is required to pay for at least 85% of the contract power throughout the period; although, hardship clauses were included for a price review if there were changes in the aluminum industry. Global economic uncertainties have postponed other proposed aluminum smelter projects. Aluminum and power intensive industry exports, however, have grown from 10.4% of total exports in 1990 to 21% in 2013 (Sæþórsdóttir and Saarinen, 2016).

Moving forward, Iceland's aluminum industry may be impacted by two particular issues. First, as a member of the EEA, Iceland was required in 1996 to deregulate its power system through an EU directive (Hreinsson, 2007). It began to address this directive with changes to its Electricity Act in 2003, when utilities were unbundled into a generating company, Landsvirkjun, as well as the partial unbundling of other utilities into separate departments for generation, sales, and distribution. The companies remain state-owned; privatization remains uncertain. Second, however, as more manufacturing and tourism industries evolve, there are questions about how these power resources will be able to address the needs of all stakeholders (Sæþórsdóttir and Saarinen, 2016). It is estimated that the aluminum industry uses roughly 83% of the energy produced in Iceland.

Emergence, transition, or continuity

The new regional economy of staples-producing regions seem to both resemble and depart from their older incarnations. These resemblances and departures arise from the continuing role of resource-producing peripheries and suppliers of raw materials to global manufacturing cores. They also arise from two post-1980 processes – the technical and process shift from Fordist to more Flexible modes of production, and the geographic shift in the global manufacturing core from the United States to Asia.

When we look from the perspective of Staples theory, we can identify that large industrial interests are continuing to 'buy-in' to natural resource sectors to meet their supply chain needs. Relative shifts over time have included changes in the participation of national versus international firms, and by an increasing participation of Asian-based investment. When it comes to questions of local benefits, these continue to be focused upon local employment while profits flow out of the resource-producing regions to capital and to the state. For those regions with Aboriginal or Indigenous populations, there has also been increased attention to revenue sharing and impact–benefit agreements. In many cases, the impetus for these types of agreements is the underlying (though many times as yet unsettled) treaty and legal claims that Aboriginal / Indigenous peoples hold.

Questions arising from EEG lead us to identify that indeed changes are occurring in these resource-producing regions. In addition to the scope and scale of activities, we note closure and dramatic restructuring of older industries. There are also changes in the ownership structure and changes in the type of natural resource leading regional development, especially where they fit nicely into the pre-existing regional institutional framework. These changes appear to be both incremental as well as dramatic in their application.

At a more general level, points of similarity and resemblance to the older political economies of resource-producing areas are many. To start, the regions are still very much embedded in a staples economy that continues to focus on the production of low-value raw materials as industrial inputs for more advanced economies outside the region and hemisphere. Despite calls for diversification, there are limited examples of such within regions that continue along a development path, characterized by a narrow range of products, the demand and price for which are subject to wide fluctuations.

A second feature of similarity and resemblance comes from the continuing dependence on large industrial capital to undertake new projects. As new projects still require global economic players building and managing their development, construction, and operations, this is likely to perpetuate the truncated development aspects of a region's staples economy.

Points of departure are also many. Some are more nuanced and others more dramatic. In terms of nuanced differences, there is the shift in target markets from continental North America to Asia. Among the more dramatic has been the replacement of what had been the only remaining tangible local benefit from resource development – local employment. The use of temporary foreign workers and fly-in / fly-out employment rosters can significantly reduce local and regional benefits of resource commodity production (Markey and Heisler, 2011). There is the potential that these resource-producing regions will realize very few benefits while bearing all of the long-term social, cultural, and ecological costs of non-renewable resource extraction.

As a counter, the increasing participation of Aboriginal / Indigenous peoples in dialogue over development proposals may see more joint partnerships, revenue sharing, and impact / benefit agreements come into place. This would reverse some of the outward flow of benefits (in the form of employment and cash). The question remains, however, whether such cash payments can ever compensate or account for the impact of non-renewable resource extraction in traditional territories.

With significant historical investment in industrial development, and potentially significant new industrial investment, have resource-dependent regions changed? Do we see the emergence of a new and diversified economy, do we see transitions within the regional economy, or is there a continuity of its traditional focus and dependence? Without drawing very deep from the insights of a Marxist critique of the concrete and the abstract when interpreting economic restructuring, it seems clear that aspects of the concrete have changed. In the language of EEG, the shifts and changes in orientation or focus within the general trajectory of path dependency are incremental at best. However, when we consider, even partially, the abstract, it is equally clear that the changes described represent a continuity of resource-dependence. This leaves in place the imperatives of a capitalist economy with firms motivated for profit looking for competitive advantage. These firms engage in the extraction of staples resources and ship minimally processed commodities into a global economic system. Within that system, increasing competition from other producer regions around the world continues to put general downward pressure on commodity prices which close the loop back to the firm which then needs to find cost-cutting ways to protect market share and profitability.

Closing

This chapter is the last of our background chapters. It completes the introduction of a capitalist economy that is structured, and which functions, within a global economic system. In this chapter, we introduced a suite of theoretical frameworks that are useful for interpreting the specific political economy of resource-dependent regions in developed economies. Staples theory, evolutionary economic geography, and institutionalism each contribute to the understanding of resource development and the path dependency that it creates.

Staples theory continues to be a vital and relevant construct given that the principles of the relationship between resource-dependent places and regions in developed economies to the rest of the advanced economic system remain the same. Challenges that result from dependency and truncated development persist, and leave such places and regions vulnerable to changes in markets. The involvement of transnational firms, though not a new phenomenon, has certainly continued and their activities are assisted by the technological and communications revolutions of the past 30 years.

Evolutionary economic geography has challenged the notions of path dependence that were often taken from the Staples theory approach and has asked the question of whether change is possible. As with our previous critique of restructuring, it does appear that change is not only possible, but that it is a requirement of firms in a capitalist economy if they are to stay competitive and profitable. When EEG asks questions about whether incremental or evolutionary change is enough to break from past path dependence, the evidence from resource-dependent places and regions is mixed. If there is a catastrophic event such as resource exhaustion or complete closure, there will of course be a break. If resource exporting remains the key economic driver, the processes of change are rather more about small adjustments within a still-robust path dependence.

One of the key tools for supporting continued path dependence is the array of structures and institutions, as well as the norms and habits, already in place. Each of these not only facilitates, but also has a vested interest in perpetuating, the contemporary political economy. Whether it is the allocation of resource use rights, the nature of the economic infrastructure and the skill set of the workforce, the organization of political and social power, or other specifics of the techno-economic paradigm, institutionalism has been a useful addition to our conceptual tools for understanding resource-dependent places and regions.

The second half of the chapter was devoted to a number of historically informed 'stories' about the development and changes that have been woven through resource regions. With the background provided in this chapter, and its companion chapters in this part of the book, we can now start to explore some of the more specific elements of the political economy of resource-dependent places and regions. How are the trends of change in public policy impacting these places and regions? How is the state being impacted by, and how is it responding to, the accelerating pace of globalization? How are communities and local governments? It is to these topics that the next part of the book is devoted.

Part III
Mobilizing through institutions

Introduction

The preceding chapters set the foundations for understanding change in resource-dependent remote, rural, and small town places and regions within developed economies. They set out the nature and underlying imperatives of the capitalist economy, that accelerating globalization of the capital economic system creates an uneven distribution of costs and benefits, and that for rural or remote resource-dependent (or staples-producing) regions, there is a tendency to become locked-in to a natural resource-development pathway. Change over the past 40 years has also been uneven. Some resource-dependent regions have diversified, others collapsed economically, and still more have witnessed only incremental change as they continue to struggle with waves of economic booms and busts.

This part of the book turns our attention to the present and to the future. It takes our understanding of resource-dependent places and regions and asks how are the trajectories of change working through the institutions that play key roles in both maintaining and changing these places and regions. The individual chapters in this part also look forward with respect to the implications of past and present trends on these different institutional actors – and on the resource-dependent places and regions at the heart of our concern. Through the following five chapters, topics introduced in the previous part of the book are mobilized through these key institutional actors.

Chapter 6 pulls together some of the most relevant institutional trends. The chapter opens with a review of the concept of institutions and institutionalism. At the core of the chapter, however, is the transition from a post-war Keynesian political economy framework to a Neoliberal political economy framework. This transition has had wide and significant impacts, not only in terms of changes in state policies or approaches to resource development, or rural development more generally, but it has also reshaped expectations across numbers of sectors. When coupled with the speeding up of the global economy facilitated by new information technologies, this transition generally puts vulnerable places into even more vulnerable positions.

Chapter 7 turns attention to the role of small businesses and large corporations. In resource-dependent regions, these two ends of the private enterprise spectrum seem to dominate. Resource industries have, since the end of the Second World War, been the purview of large international capital. The scope and scale of investments, the connections to international supply and marketing chains, and increasing challenges around efficiency of production to remain competitive and profitable in the face of increasing competition from low-cost producing regions has supported the continued involvement of transnational corporations.

At the other end of the spectrum, the service and supply businesses in resource-dependent regions tend to be branch outlets of chains (banks, pharmacies, equipment dealers), which have a local owner or franchisee, or they are locally owned small businesses. These economic actors are key boosters of the local / regional industrial giants and have been voices for the adoption of Neoliberal political economy approaches that reduce fiscal and regulatory burdens on industry, while at the same time reducing similar burdens on small businesses. A further consequence has been the reduction of general government services to the local population, something which opens up local entrepreneurial opportunity. However, these local economic actors are also in a vulnerable position with their dependence on the direct or indirect economic spinoffs of the resource industry(ies) within the region. Industry downsizing, production curtailments, or closure has immediate and dramatic impacts on these local businesses.

Chapter 8 focuses on senior governments. Depending upon the political system, there may be only a national government and then local governments, or there may be a national government, an intermediate tier of provincial or 'state' (using the US model and nomenclature) government, and then local governments. While the next chapters turn attention to local governments and local communities, this one is concerned about the impacts of changing trends and trajectories on senior governments. Central in this chapter is the shift from a Keynesian to a Neoliberal political economy approach to public policy and its impacts on resource and rural community development. The chapter also explores how the dialogue on taxation and government spending has impacted its capacity to support community and economic development as well as local and regional transformation. Next, the chapter reviews some of the internal expectations and external pressures that limit the flexibility of both policy and governance approaches more generally.

Chapter 9 focuses upon place-based communities and local government. This is a critical institutional environment, for it is in the small communities of the resource-dependent frontier of developed economies that globalization and national trends in both the economy and public policy 'hit the ground'. The consequences of decisions in both economic and policy spheres are also now felt so much more quickly than in the past. The chapter opens with a discussion of how we define and conceptualize place-based communities and their role and 'situatedness' in the global capitalist economy. Next, the chapter introduces the concept of local government and reviews the impacts and implications for them of changes at the senior government level. With increasing responsibility for community and economic development planning, and for maintaining a readiness to respond to economic transition, local governments typically lack the fiscal and policy tools to fulfil this growing component of their mandates. The chapter closes with a discussion of new regionalism. In a world where big business is becoming bigger in order to compete, small places need to find ways to have an impact in policy and investment decisions. They need to find a way to come together with their neighbours around issues of mutual interest. Regional collaboration is neither new nor easy, and the chapter reviews the opportunities and the barriers to regional collaboration in a Neoliberal political economy.

But the institutional 'thickness' of place-based communities goes well beyond local government. For that reason, Chapter 10 introduces and focuses upon local service delivery and the increasing role of the voluntary sector in the delivery of those services. Both of these topics, service delivery and the voluntary sector, are under stress in resource-dependent small communities. Services are being closed or regionalized by senior governments looking to reduce their costs and budgetary deficits, or by private firms struggling with profitability. Into the gap we find that the local voluntary sector often steps, but in places with an aging or declining population base, the increasing burden of tasks runs the very real risk of burning out the local volunteer base.

Chapter 11 changes the scale of our institutional focus from firms, organizations, or groups to that of the individual. After introducing this scale of inquiry, most of the chapter focuses on the implications of restructuring for workers and their families. The relative stability of the Fordist mode of production is described together with its implications for workers, families, and communities. The transition to a more Flexible mode of production is then described by drawing out the impacts and implications for the same set of worker, family, and local community interests. In a world with increasingly mobile capital, we are witnessing an increase in mobile workers – people whose workplace and residence is separated by a significant distance. Some commute electronically via computers and other information technology, others commute by car or plane for long shift rotations away from home. These changes have significant implications for workers, families, and communities.

6   Institutional trends

Introduction

This part of the book focuses on a select set of institutions that are key to the development and redevelopment of resource-dependent remote, rural, and small town places and regions within developed economies. To set the foundation for this part, this chapter opens by providing an introduction to the concept of institutions and to the theoretical debates around understanding the role of institutions in community and economic development and change. It then turns attention to the significant 20th-century transition from Keynesian public policy approaches to Neoliberal public policy approaches. This transition in approaches impacts each of the institutions described in the following five chapters: business / industry, the state, local government, civil society, and individuals. Given earlier discussions, within a number of theoretical frameworks, about the central role that institutional structures play in maintaining and reproducing the trajectories of power and uneven development, understanding institutions and institutionalism is important.

Institutions and institutionalism

Before turning to a discussion of specific approaches to understanding institutionalism, we first need to introduce the general concept of 'institutions'. As noted in the previous chapters, most theoretical explanations for the structure of the economy generally, or for the structure of regional resource-dependent economies specifically, recognize that supporting these structures is a suite of organizational and institutional arrangements. Such institutional arrangements are necessary for the management and reproduction of the particular political-economic structure being described.

Within some theoretical explanations, institutional frameworks are viewed as being relatively 'rigid' and committed to the reproduction of the status quo. In these cases, they are not simply the locus of management tasks, but they also exercise power and manage conflicts to ensure the continuance of the particular political-economic structure. Within other theoretical frameworks, however, institutions and institutional frameworks are understood as a driving force behind change. For still other theoretical frameworks, institutions and institutional frameworks are a co-opted force employed through the will of another 'driver', such as capitalism, which seeks to set particular political-economic structures into place.

In his work on defining institutionalism, Amin (2009, p. 386) writes that it is a "term with many meanings in the social sciences, all intended to signal the varied ways in which institutions structure social life in time and space". He goes on to write that it "also seeks to interpret structure in terms of historically and socially embedded institutions, seen to evolve slowly, often unpredictably and sometimes inefficiently" (Amin, 2009, p. 386).

Both Hayter and Innis draw upon institutions and institutionalism to provide structural support for their analyses. For Hayter's (2008) example of industrial forestry, the particular arrangements of government policy and regulation; corporate ownership and management structure; global market conditions; and the available harvesting, processing, and transportation technologies together form discrete techno-economic paradigms (see more on techno-economic paradigms below). Allowing for evolutionary change over time, such paradigms shift ever so slightly over time as changes occur within these different arrangements.

Looking back to the lessons from Staples theory, Innis recognized that change was indeed part of the resource frontier and that such change was dependent upon the instability inherent in the three identified central forces of geography, technology, and institutional arrangements. All three were subject to change over time and this creates ongoing disequilibriums and challenges to the status quo. Reminiscent of Harvey's (1990) critique of restructuring and flexible accumulation, while the structure of staples dependence remained (the 'abstract'), the particular form and focus of that staples economy (the 'concrete') was subject to shifts and change over time. New resources, new markets, new corporate players, new business or political elites, changes in the cost of energy, changes to critical transportation infrastructures, and different public policy approaches are some of the many ways to introduce change into the institutional structure. In Canada for example, despite over 100 years of changes in the previously listed topics, resource-dependence remains the central characteristic of the economy for many regions, and indeed for many provinces. The same can be said in Australia and numbers of other developed economies. Issues of dependence and truncated development within a global economic system remain plainly obvious, given the degree to which downturns in the demand for natural resource commodities wreak uncertainty, instability, and sometimes even havoc in these economies.

Similarly, evolutionary economic geography (EEG) recognized the critical role of institutions and institutional arrangements, and it also recognized that rigidity in such structures should be replaced instead with an understanding of slow-moving and deliberate transformation. In this case, the understanding of institutions embodies both the notion of change as well as the ongoing role of those institutions in producing, managing, and reproducing the political-economic structure. Drawing on chapters from earlier in the book, we must evaluate in whose interests these slow-moving and deliberate transformations are being managed. Commensurate with the reduced labour opportunities and tax benefits found within resource-dependent places and regions, we can see that these adjustments and transformations are not serving those people and places, but are instead serving the needs of industrial capital. Less clear, however, is the degree to which they may be serving the state – which relies upon the revenues from natural resource industries and natural resource exports, but which is also experiencing diminishing relative revenues from those same industries over time.

Institutionalism recognizes that the forms and structures of regulation and organization follow both formal and informal practices. As noted in the previous chapter, formal regulation includes laws, policies, standards, rules, etc., while informal practices include habits, norms, routines of behaviour, customs, etc. These formal and informal practices can occur within companies, within governments, within communities, and more generally within society. Organizations and institutions employ both formal and informal practices to maintain their efficient functioning and their seamless fit with the techno-economic paradigm and the particular political economy structure within which they operate. As with EEG, institutionalism also allows for the purposeful transformation of both the structures and the practices of regulation and organization.

While other theoretical frameworks use the label of 'institutionalism' to describe the array of structures that keep a political-economic system in place, for Harvey (1990, p. 121), the regime of accumulation refers to the actions and behaviours of "all kinds of individuals – capitalists, workers, state employees, financiers, and all manner of other political-economic agents – into some kind of configuration that will keep the regime of accumulation functioning". In turn, a regime of accumulation needs a manifest suite of "norms, habits, laws, regulating networks and so on that ensure the unity of the [production] process ... [This body of] rules and social processes is called the *mode of regulation*" (Harvey, 1990, p. 122).

Linking with the structural explanations of a Marxist political economy and writers such as Harvey, Amin (2009, p. 386) argues that institutionalism very often "rejects actor-centred approaches that stress individual human intention and will". This, of course, circles back to our earlier observations about theoretical debate, especially those that attribute intentionality to larger structures such as the capitalist economy versus those that allocate more responsibility to human agency and individual human intentions or actions. As argued in Chapters 2 and 3, we favour an intermediate position that recognizes the imperatives, opportunities, and barriers embedded within broad institutional structures, but within which there is limited and bounded room for individual human agency and decision making.

As noted in Chapter 5, how these institutional arrangements are structured highlights yet again the importance of focusing upon communities and regions as the meeting point for both global and local forces. With our interest in resource-dependent places and regions within developed economies, historically produced institutional structures remain deeply embedded and reticent to change. House (1999) highlighted the role that one group of institutional actors can have on maintaining both structures of power and the established political economy. Between 1989 and 1996, House served as the chair of an economic development agency called the Economic Recovery Commission in the province of Newfoundland and Labrador, Canada. The Commission was charged with identifying an integrated and balanced approach to both social and economic development in a province that had significant economic transition challenges. In reflecting on why the work of the Commission failed to yield transformative change, House (1999, p. x) writes at length

> about the barriers and impediments that frustrate or thwart attempts of economic renewal in Newfoundland and Labrador. The work of the Economic Recovery Commission was partly undermined by the workings of government itself. The provincial bureaucracy, led by a powerful group of senior public servants that I refer to as the Old Guard, is a control apparatus that systematically resist change, undermines the innovative efforts of agencies such as the ERC, and is highly successful at inducing successive political élites (premiers and ministers), mainly unwittingly, to support its approach. In a society that has become highly dependent on government, the Old Guard within the provincial bureaucracy (abetted on some issues, such as income security reform, by the federal bureaucracy) is a powerful conservative force that perpetuates dependency.

In his book, House (1999) not only describes the work of the Commission, but he also goes on to detail the various ways and processes by which political and bureaucratic elites can effectively stifle change. In some cases, these strategies are grounded in the very human responses of jealousy or protection of privilege or power; at other times they are grounded in ideology. In outlining the various mechanisms for obstructing change, House (1999) describes everything from passive resistance to the control of information and information flow, to subtle forms of discrediting both people and processes, to exaggerating minor failings or mistakes. Further obstructing actions and mechanisms occur when government seeks to implement change. These include starving initiatives of funds, limiting them to pilots and policy studies, and rendering them nearly invisible within the managing structure of the government. Intended or otherwise, these institutional responses serve to fulfil expectations from the theoretical literature about the role of institutions in producing and reproducing the status quo within the dominant political economy.

But the role of institutions in supporting or thwarting change is the subject of much debate. In describing the bureaucratic institutional structures that might facilitate or impede public policy change, Swarts (2013, p. 137) adds to

House's observations about powerful elites but takes a different tone in recognizing that such elites can sometimes be "harnessed" by a "more consensual, consultative, and deliberative" political structure.

In this case, the more 'open' the structures of government, the more venues there are for opposition groups or interests to intercede, delay, sway public opinion, or even hijack debates. The more closed the structures of government the more it might be directed to a purpose. In the discussions below about the tactics of advocates for Neoliberal policy change, we can see that they effectively used more closed government structures and powerful majority governments (and charismatic government leaders) to force their agenda for public policy change.

Techno-economic paradigms

How then, can we amalgamate the different approaches to, and understandings of, institutionalism? One way is via the notion of techno-economic paradigms. As introduced in Chapter 5, techno-economic paradigms summarize the arrangements at any particular time and place of the technology (understood at a very general level such as coal-steam or silicon chip-computer technologies) driving economic production, and of the social and political structures needed to support that technology arrangement. As such, not only are techno-economic paradigms time- and place-specific, but they also embody an imperative for change through the competitive processes of the economic market working over time.

Our summary of techno-economic paradigms draws mainly from the work of Roger Hayter. This is not only because of the role techno-economic paradigms have played in his research, but also because his research focuses specifically on the restructuring of natural resource industries. As described by Hayter (2000, p. 5),

> Emerging techno-economic paradigms centre on major new technologies that have pervasive effects throughout the economy, creating new forms of production and engineering principles, new organizational arrangements, and government economic and social policies ... The rationale for a shift in the techno-economic paradigm occurs when new forms of production, technology, and engineering principles (and institutional arrangements) offer substantial improvements in productivity over prevailing systems and ways of thinking, especially if the benefits of the latter have more or less 'played themselves out'.

Such rationales for changes and shifts not only impact the technology being employed for the organization of the production process, but they have wider implications. As described by Hayter (2000, p. 8),

> The technological changes associated with each techno-economic paradigm are paralleled by new regimes of regulation and institutional

innovation involving new forms of business organization, research and development (R&D), and labour relations, as well as macroeconomic government policy initiatives and even new ways of organizing the international economy.

These changes and shifts through the transition to new techno-economic paradigms are thus both extensive and pervasive. They create both crisis as well as opportunity, as old systems and institutions (structures as well as the organization of both economic and political power) are dismantled and new ones created.

In his analysis of transition in the British Columbia forest industry that started in the late 1970s, Hayter (2000) does not focus upon the policy transition to Neoliberalism, but instead focuses upon transition in techno-economic paradigms. In this case, the focus is upon the transition from the Fordist techno-economic paradigm to the information and communication techno-economic paradigm. In describing that emergent information and communication techno-economic paradigm, Hayter (2000, p. 12), summarizes a number of key attributes from the literature. These include:

- The use of new technologies so that more flexible production can make more efficient "use of materials, space, and workers".
- A continuing important role for multinational corporations, but flexible production now creates niche opportunities for small and medium-sized firms.
- Networking and sub-contracting alliances that tie together larger and smaller corporations through integrated and real-time strategic production management.
- Drives for efficiency that mean there has been a flattening of corporate hierarchies and a reduction of the management middle class.
- The application of flexibility principles to labour that now supports the increasing use of casual and part-time workers.
- That there has been a reduction in the bargaining power of unions and other forms of organized labour that otherwise would impede the 'freedom' of capital flexibility.
- And that "All employees are expected to contribute to innovation, R&D activities are closely integrated manufacturing activities".

The core themes of this list link back to our earlier discussions of restructuring and flexible production.

Changes in approach

In working to situate the institutional approaches that have supported resource-dependent economies, we look at two broad characterizations. The first, which emerged out of the political-economic challenges of both the

Great Depression and the immediate aftermath of the Second World War, has been called Keynesianism. It is named after John Maynard Keynes, the economist who postulated its central tenets of state intervention in the market to control excesses and to smooth the economy's most dramatic fluctuations.

The second, which came to prominence beginning in the late 1970s and 1980s, is called Neoliberalism. Deriving from its 'liberal' root, the label identifies a focus upon individualism and individual freedoms. When extended to the economy, the focus is upon a reduction of state intervention and an enhancement of the 'freedom' of market processes. Both of these economic theories came to be institutionalized within the formal and informal practices of organizations and regulatory structures.

Keynesian approaches

State intervention to dampen business cycles was part of the policy management response to controlling the excesses of the market that had led to the catastrophic Great Depression (excesses that are being recreated from time to time it seems – most recently the 2008 stock market collapse and resulting global economic recession). But the experience of the Great Depression added a second key element to a Keynesian public policy approach – that being the role of government intervention, often through major investment and infrastructure projects, in stimulating economic activity and job growth. As described by Harvey (2005, p. 10), this particular political economic focus included:

> an acceptance that the state should focus on full employment, economic growth, and the welfare of its citizens, and that state power should be freely deployed, alongside of or, if necessary, intervening in or even substituting for market processes to achieve these ends. Fiscal and monetary policies usually dubbed 'Keynesian' were widely deployed to dampen business cycles and to ensure reasonably full employment.

The adoption of a Keynesian public-policy approach, especially its economic stimulus component, coincided in the immediate Second World War period with the need to address two imperatives. The first was employment, and how to redeploy the millions of soldiers returning from the war effort – a great many of whom just six or seven years earlier had been the unemployed masses of the Great Depression. The second was how to address the massive infrastructure deficits that would be required to allow the wartime experience of massive industrial production and global production supply and distribution chains to transition into the efficient post-war production of consumer goods. Infrastructure needs included upgrading port, rail, and road infrastructure. It also involved expanding the new critical infrastructure of airports to support the emergent air transportation industry. In both Canada and

United States, it included the creation and expansion of cross-continental national highway systems. At more local levels, there was the massive infrastructure needed to support affordable suburban housing for the baby-boom generation.

It is difficult to recall sometimes, but the investment mindset of a Keynesian public policy approach not only made practical sense given the experience of the preceding decades, but it also addressed critical political questions. At the close of the Second World War, the global contest between communism and democracy (and their surrogates in the Soviet Union and the United States) was just beginning. Addressing infrastructure needs through state-led investment not only allowed the economy to transition and develop, but it also absorbed and re-deployed labour. For historians such as Jackson (1985), the creation of the post-war suburban landscapes and associated cultures of consumption, not only stimulated the economy, but it also acted to bind citizens to the economy through homeownership, 25-year mortgage payment schedules, and the pursuit of a 'better life'.

In many resource-dependent remote, rural, and small town regions within the developed economies of the day, the dual Keynesian actions of stimulating employment through economic infrastructure investment renewed, modernized, and expanded these regions. The case of BC's industrial resource policy after the Second World War most certainly seems to fit with the balanced approach suggested via Keynesianism. Building on the recommendations of the province's Post-War Rehabilitation Council, investment to 'industrialize' BC's resource economy potential was to be the goal. The Council's report,

> presented a balanced view of the role of government, relative to private industry, in the development process. It rejected a policy of 'laissez-faire for one of intelligent, positive action' … Such intelligent and positive action required actions that are planned, coordinated, and conducted in cooperation with the other sectors of society and other government authorities.
>
> (Markey et al., 2012, p. 99)

In describing BC's leader through those post-war years, W.A.C. Bennett, who was also a member of that earlier Post-War Rehabilitation Council, and his approach, Mitchell (1983, p. 259) writes:

> Bennett's perception of the relationship between private enterprise and provincial development was highly pragmatic. When private interests failed to co-operate with his vision, he never hesitated to fill an economic void through public-sector participation. Apparently he did not push public enterprise for its own sake, for he steadfastly acknowledged and approved private enterprise as the foundation of the BC economy. But neither did he shrink from taking direct government action if he felt that private industry would not act … A fitting working motto for his

government and his era would be intervention if necessary, but not necessarily intervention.

In a recent edited volume looking at long-run processes of change in resource-dependent rural and small town regions (Halseth, 2017), each of the authors describe trends in state policy and investment similar to that described above for BC. Tykkyläinen et al. (2017) write about Finland and the northern expansion of industrial forestry. In that case, the need for absorbing excess labour and state investments in critical transportation and service infrastructure to support industrial access to resources and then to move products to market very much mirrors the BC case. In a region of New South Wales, Australia, Argent (2017a) focuses on the growth and expansion of an agricultural economic base that also depended upon state policy and investment. As in the BC case, a 'settler' economy and landscape required not only state investment in infrastructure, but also significant policy attention to labour and quality working environments in order to attract people to the region and its industries. Writing about the southern region of New Zealand, Connelly and Nel (2017a) are able to include examples from both agriculture and mining (especially coal mining). As with the Australian case, economic growth and stability in the immediate post-war period required state infrastructure investment and supportive public policy.

Neoliberal approaches

In his review and critique of Neoliberalism, Harvey (2005, p. 2) begins by noting how Neoliberalism is:

> in the first instance a theory of political economic practices that proposes that human well-being can best be advanced by liberating individual entrepreneurial freedoms and skills within an institutional framework characterized by strong private property rights, free markets, and free trade. The role of the state is to create and preserve an institutional framework appropriate to such practices.

As a logical extension, Neoliberalism "holds that the social good will be maximized by maximizing the reach and frequency of market transactions, and it seeks to bring all human action into the domain of the market" (Harvey, 2005, p. 3). For Tennberg et al. (2014, p. 42), therefore, "the role of the neoliberal state is to secure proper conditions for markets to function".

The debate supporting a replacement of a Keynesian political economy policy approach with a Neoliberal political economy policy approach has been directed and forceful. Since the 1970s at least, significant global economic players became unwavering advocates. In their critique of one such key global institutional actor, the World Bank, Waeyenberge et al. (2011, p. 6) note how the Bank was instrumental in promoting the neo-liberal perspectives on develop-ment that came to dominate the agenda of many international develop-ment actors during the 1980s. ... [including] eliminating all obstacles to a 'perfect market' as the presumed optimal path to growth. This implied an emphasis on 'fiscal discipline', curtailment of government subsidies, interest rate liberalization, trade liberalization, privatisation and deregulation.

Given the role of key global economic actors such as the World Bank in assisting with economic transition in developing states, Neoliberal policy approaches are often extended and embedded by making them a condition of financial aid or loans support. This approach and 'lever', by which Neo-loberal policy was being extended in developing states, was more recently adapted for use in developed economies that were struggling as a result of the 2008 global economic recession – especially those of the European Union's southern member states of Greece, Italy, Spain, and Portugal. The example of Greece is telling as it involved a range of Neoliberal policy actors. With eco-nomic catastrophe looming, financial assistance to the Greek state came with conditions around the aggressive dismantling of the welfare state and its social democratic supports. This attack on the expenses of government per-sisted despite evidence from the revenue side that wealthy Greek corporations and individuals were not paying taxes.

With this theoretical perspective, the 'project' of Neoliberalism needed to target the dominant policy approach of the day in order to replace it. As described by Harvey (2005, pp. 20–21), Neoliberal theorists in the post-war period

held to Adam Smith's view that the hidden hand of the market was the best device for mobilizing even the basest of human instincts such as gluttony, greed, and a desire for wealth and power for the benefit of all. Neoliberal doctrine was therefore deeply opposed to state interventionist theories, such as those of John Maynard Keynes, which rose to promi-nence in the 1930s in response to the Great Depression. Many policy-makers after the Second World War looked to Keynesian theory to guide them as they sought to keep the business cycle and recessions under control.

That replacement process needed to be wide ranging as the political economy of a socio-economic system infiltrates all aspects of the economy and society, as well as the regulatory norms, processes, and institutions that manage and protect that political economy.

As Swarts (2013, p. 3) describes the transition:

over the course of the 1980s and 1990s, in countries governed by parties of both the right and left, traditional policies of Keynesian demand management and active state economic intervention gave way to policies – variously dubbed 'neoliberal,' 'economically rationalist,' and 'new right' – privileging the role of the deregulated, 'free' markets.

The extent of the political and policy replacement process has been remarkable (Peck, 2010). Despite that, 30 years of Neoliberal practice has dramatically reinforced the political and economic power of an increasingly small elite, this new political-economic imagination continues to hold sway.

A definition

A number of writers draw upon the work of Hay (2007) in framing a broader and more comprehensive definition of Neoliberalism. Hay (2007, p. 97) outlines a series of eight central 'tenets' for Neoliberalism:

1   A confidence in the market as an efficient mechanism for the allocation of scarce resources.
2   A belief in the desirability of a global regime of free trade and free capital mobility.
3   A belief in the desirability, all things being equal, of a limited and non-interventionist role for the state.
4   A conception of the state as a facilitator and custodian, rather than a substitute for market mechanisms.
5   A defence of individual liberty.
6   A commitment to the removal of those welfare benefits that might be seen as disincentives to market participation (in short, a subordination of the principles of social justice to those of perceived economic imperatives).
7   A defence of labour market flexibility and the promotion of and nurturing of cost-effectiveness.
8   A confidence in the use of private finance in public projects and, more generally, in the allocative efficiency of market and quasi-market mechanisms in the provision of public goods.

As described by Swarts (2013, p. 4), the implications were dramatic and sweeping:

> Government economic intervention, market regulation, tariff protection, and labour-friendly policies were drastically scaled back. Deregulation, attempts to control government expenditure, reduce debt loads, eliminate inflation, free up world trade, and reduce the influence of traditionally powerful trade unions, became widespread.

In New Zealand, Connelly and Nels (2017c, p. 116) described the post-war policy landscape:

Prior to the 1980s, Keynesian thinking dominated state economic policy in New Zealand. State control became a dominant feature of the economy after the Great Depression in the 1930s and was marked by high levels of state ownership of industry and resources, spatial interventions, and the regulation of individuals and firms ... Trade and financial controls, centralized bargaining, state subsidies, control of strategic industries, and state welfare were hallmarks of the era.

This changed in 1985, when a Labour government introduced the first of a wide sweep of policy changes. Following along from the points noted above by Hay and Swarts, in time these policy changes included that:

State-owned enterprises and control boards were replaced by policies of corporatization and then privatization, state services were trimmed in almost all sectors, and state enterprises were cut ... In parallel, industrial and agricultural subsidies and control boards were suspended, the currency was allowed to float, and more flexible labour market conditions, which did away with collective bargaining, were introduced.

(Connelly and Nels, 2017c, p. 117)

Ioris (2016) recently published a political economy analysis of Neoliberal agribusiness. Recognizing our earlier description of the imperatives of capital and capitalism, Ioris (2016, p. 85) writes about how contemporary agriculture has been reshaped and subjected

to the imperatives of flexible accumulation, market globalization and the systematic concealment of class-based tensions. The intricacies of global agri-food activities today are, at once, product and also co-producer of the dominant modernization of capitalism according to the discourse and strategies of neoliberalism.

Following Swarts, these strategies constituted "an assertive programme aimed at dislodging the politico-economic approaches adopted before the 1980s" (Ioris, 2016, p. 86). But the application of a theoretical and ideological Neoliberal public policy and market approach encounters a critical contradiction in practice in that "Neoliberalised agriculture has also evolved through an inconsistent argument about the virtues of free market transactions, whereas there are simultaneous calls for sustained state interventions expected to regulate price oscillations and avoid over production" (Ioris, 2016, p. 86).

A policy transition

In his study of the Keynesian to Neoliberal policy transition in Britain, Canada, Australia, and New Zealand, Swarts (2013) argues that these four countries experienced a dynamic social reconstruction via both political and economic change. This social reconstruction was "promoted by elites as part of a strategy to reset the basic parameters, expectations, and shared norms of the relationship between the state, society, and the (inter-)national economy" (Swarts, 2013, p. 5).

Swarts (2013) makes three main arguments about how elites undertook to mobilize and advance their interests within this social reconstruction process. The first includes an emphasis on not just persuasion, but also coercion. Second, these instruments of persuasion and coercion were deployed together with various material incentives such that they helped induce conformity within the new economic structures and that "experience and habituation led the new ideological and policy paradigm to become 'taken for granted'" (Swarts, 2013, p. 6). Third, Swarts (2013, p. 6) employs an institutional approach to highlight how "the opportunities and constraints presented by each country's structure of political institutions" was employed in the social reconstruction process and was deployed by particular political elites as part of a strategic process to secure political advantage and power to pursue additional change.

But how can we understand the individual components that structure the process of dramatic change in policy approaches? As described above, the end product, or goals, of the policy approach transformation have been reasonably well articulated in hindsight. Also, the tactical approaches used by elites, including both persuasion and coercion have been reasonably well described in the literature. At this point, however, it is worth spending a moment looking at some of the components of that process of change in the policy approach underscoring the political economy. To this question, Swarts (2013) introduces four building blocks. The first of these building blocks involves structural change. As noted in earlier chapters, increasing globalization and the intense pressures from competition and the exhaustion of the benefits that could be derived from a Fordist production model were marking a period of intense structural change in the economy. This created pressures and imperatives on capital to address its internal crises of profitability. It also generated crises on the part of the state, which was seeking to maintain its critical revenue flows from those industrial production sources, while at the same time balancing decreased revenue flows by managing and curtailing its own expenditures. Expenditures on social services had been a significant and growing portion of state budgets in the post-war era. The size of these social service expenditure envelopes within state budgets made them easy targets for systematic and long-term incremental reductions.

Policy change was the tool through which the transitional supports for both capital and the processes by which states managed their own revenue and expenditure activities was mobilized. Deregulation, together with efforts to reduce operating costs for industry became one cornerstone of policy change. Reorientation of state attention and expenditures away from social service provision and the contraction of welfare supports became another policy change cornerstone. In turn, both of these were accompanied by an unrelenting Neoliberal discourse on how state policy can become more facilitative rather than directive, and how it can promote individual (at the corporate and citizen level) freedom. The coincident rise of personal computers and the internet also provided a convenient way for the state to support these incremental reductions and curtailments in the costs of direct services delivery by putting more information and applications processes online under the guise of freeing up access – never mind that home computers and access to the internet continue to be limited by economic barriers for some of the most vulnerable segments of society.

Both of these first two dynamic components of structural change and policy behaviour needed, however, a wider framework to guide change. The coincident rise of a Neoliberal discourse in economics, and later in public policy, provided such a conceptual frame. Neoliberal ideology provided both a logic and a specific set of targets for action. Comprising an ideological faith in the free market, the Neoliberal discourse targeted its actions towards a systematic dismantling of the Keynesian public policy approach. Caveats to this dismantling include that successive crises of capital in the Neoliberal era have been increasingly addressed by aggressive state intervention and support of failing capital enterprises. Finally, these shifts across components helped to communicate through society and the economy a new image of the political economy. While a Marxist critique would suggest that this was a rapid expansion of 'false consciousness' among the workers, people from across society began to inculcate a sense of opportunity and free enterprise despite that the overwhelming proportion did not realize the direct benefits – in fact, in resource-producing regions they bore the brunt of closure and capital flight.

In terms of the process of debate itself, Swarts (2013) identified four basic arguments that dominate the persuasion, coercion, and rhetoric behind the Neoliberal policy transition. The first of these arguments involves the need to cultivate a sense of economic crisis – a crisis of sufficient magnitude so as to demand a new and alternative policy approach. Second, arguments need to link any selected transitory economic crisis to a longer-term process of national decline that supposedly demonstrates the failure of the former policy regime. The third is to identify through select international comparisons situations where the state is now in an uncompetitive position relative to the selected examples and thus in need of significant public policy revamping. Finally, Swarts (2013, pp. 139–140) identifies:

> the rhetorically coercive strategy of positing that 'there is no alternative' to the new policy orientation. Building on – and consciously contributing to – the sense of economic crisis discussed previously ... [proponents] ... sought to convince their audiences of the need for and appropriateness of substantial change, to promote the supposed benefits that would accrue from such changes, and to coerce hostile opponents or reluctant observers into accepting that no other alternative to their plans existed. It was through this conscious, strategic interweaving of these arguments that opponents to neoliberalism were largely marginalized.

Individuals and intervention

With respect to the role of the state, Neoliberal theory suggests a diminished and non-interventionist role. As described by Harvey (2005, p. 64), "According to theory, the neoliberal state should favour strong individual private property rights, the rule of law, and the institutions of freely functioning markets and free trade". Built from assumptions that the accumulation of wealth will create 'trickle-down' benefits for the rest of society, Neoliberal theory goes so far as to suggest that "the elimination of poverty (both domestically and worldwide) can best be secured through free markets and free trade" (Harvey, 2005, p. 65). Problems with the notion of trickle-down benefits are many. The first is that Neoliberal theory has not dealt with the problem of competition creating monopolistic or oligopolistic behaviour, such that a small number of firms (or individuals for that matter) can control of whole economic sectors. Second, it fails to address the behaviours that lead to market failure. The recurrence of market failures since the 1980s is evidence that the excesses of individualistic behaviours can have catastrophic impacts.

Neoliberal theory has even struggled to address some of the most basic of disjunctures and contradictions in its foundational assumptions. These include the impossibility that economic actors will all have equal access to similar information, that there will be a level playing field for the different actors, or how points of adjustment when new technologies or social relations interfere with pre-existing market structures are to be managed to avoid excessive speculation and crisis. As Harvey (2005, p. 69) argues, there are:

> some fundamental political problems within neoliberalism that need to be addressed. A contradiction arises between a seductive but alienating possessive individualism on the one hand and the desire for meaningful collective life on the other. While individuals are supposedly free to choose, they are not supposed to choose to construct strong collective institutions (such as trade unions) as opposed to weak voluntary associations (like charitable organizations). They most certainly should not choose to associate to create political parties with the aim of forcing the state to intervene in or eliminate the market … This creates the paradox of intense state interventions and government by elites and 'experts' in a world where the state is supposed not to be interventionist.

Contestation

Of course, nowhere in the literature is it suggested that the policy approach transformation from Keynesianism to Neoliberalism has gone uncontested. Conflict and debate has stretched across many fora and has occurred at many scales. To start, many of the large protests against globalization and international free trade agreements that have occurred at various world trade and climate change conferences are emblematic of both organized and grassroots mobilization against the increasing political and economic power of capital. Within states, there have been ongoing political and public policy debates associated with the retrenchment of the welfare state and the relative curtailment of social service expenditures. From time to time, in any given state context, there have been protests to raise awareness around, and demand renewed government intervention to mitigate, issues such as poverty and homelessness, job losses, industry closure as work shifts offshore, corporatization, the distribution of wealth and benefits, economic marginalization, displacement, and others. A third general area of contestation has been the environment. In some cases, issues are more general and involve topics such as climate change, global warming, species extinction, and the like. In other cases, they are more specific and target topics such as particular oil or gas developments, pipeline proposals, nuclear power plants, and what is to be done about the enormously large number of coal-fired electrical generation plants that exist already or that are proposed in particular nations. Even locally, protest and contestation sweeps across these areas from a general concern about amorphous globalization to local concerns about the maintenance of neighbourhood parks and playgrounds against development interests that seek to turn them into housing or commercial properties.

Implications

The implications of this Neoliberal policy turn for resource-dependent rural and small town places have been significant. Of course, one of the first and most significant of these impacts is economic. Many resource-producing regions within developed economies are high-cost commodity producers. They are high cost for many reasons: wages, taxation, regulation, services, etc. relative to some other regions of the globe. The process of globalization has rapidly opened up markets to the entry resource commodities from low-cost producer regions (see Halseth, 2017). In addition to ongoing fluctuations in the demand and prices for resource commodities in the global marketplace, the increased supply from low-cost producers puts further downward pressure on prices and reduces market shares for established producers.

A key response by resource industries in rural and small town regions within developed economies has been to aggressively reduce its cost of production in the face of this new competition. At the local level, most important among the strategies deployed has been a substitution of capital (especially automated and computer-controlled production machinery) for labour. This can be very successful in reducing costs per unit produced, as well as increasing output. In addition, other strategies have involved pushing costs off to local suppliers and contractors, as well as reducing operating costs from taxation. These matters will be discussed further in Chapter 7.

The first of these strategies has meant significant job losses in traditional sectors. In a recent international volume, each of the reviews of resource industry restructuring impacts on labour in Australia (Argent, 2017b),

Canada (Halseth et al., 2017b), Finland (Halonen et al., 2017), and New Zealand (Connelly and Nel, 2017b) report significant job losses over time, as industries fought to remain competitive. They also highlighted that replacement work has neither equalled the numbers nor the wages of the previous resource industry jobs.

The strategy of reducing operating costs by securing lower taxes has put pressure on local and regional government structures. Charged with more responsibility for economic development and planning, these governments have fewer relative fiscal resources to attend to the infrastructure and service needs of potential new economic activities, let alone absorb the cost of developing and implementing economic strategies themselves (Markey et al., 2008b). Connolly and Nel (2017c) draw upon the southern region of New Zealand to illustrate a common approach to such economic planning needs – that being the creation of regional development agencies or authorities. As has been well documented in the rural redevelopment literature, while such agencies or authorities may be very effective at stimulating local/regional dialogue and small-scale economic investment planning, they lack the critical fiscal and policy tools to really make a difference. These matters will be discussed further in Chapter 9.

As Neoliberal policy advocates for less state involvement, local communities have been 'enabled' or charged with taking more responsibility for their own economic development planning. One critical problem for resource-dependent rural and small town places is that the recasting of Neoliberal public policy at the state level has not yet been able to forge a coherent response to the post 1970s / 1980s restructuring of the political economy of resource-dependent regions. Their policy mindset has changed, but their tools (the structure of fiscal and policy power and authority) for supporting rural and small town places have not changed (Halseth et al., 2017b). While local communities or regions may be able to divine economic development strategies, many of the key fiscal and policy levers they will need to mobilize and realize those strategies are still held firmly by the state. Without the support of top-down public-policy, local and regional initiatives struggle. These implications for the state will be discussed further in Chapter 8.

Another critical area of impact from the Neoliberal policy turn has been the downsizing or closure of rural and small town services. As Sullivan et al. (2014) describe, such services are being degraded at a time when rural and small town places need these types of business and resident supports more than ever, so as to retain and to attract both economic activity and people. While the service delivery models of the 1950s and 1960s are no longer practical in regions with dispersed sets of small communities, rather than seeking to deploy new models that make use of innovations in organization and technology, states too often simply close or regionalize services outright using the Neoliberal-inspired metric that costs of delivery are too high when measured against urban examples. That civil society in rural and small town places, in the form of local voluntary sector groups, are now often the ones trying to fill emergent service gaps is problematic (these implications for civil society will be discussed further in Chapter 10).

Hayter and Barnes (2012) explored the implications of Neoliberalism on the forest industry economies of British Columbia, Tasmania, and New Zealand. They identified a series of limitations on the application of, and transformations resulting from, Neoliberalism. These included political limitations governed by historical and contemporary public perceptions about industry and the environment, industrial limitations arising from the corporate and labour structure of the industry and the legacies of capital investments, environmental limits connected especially to the role of environmental non-profit organizations, and finally cultural limits linked especially to Aboriginal / Indigenous rights and title.

Blending of approaches

Finally, there is one inherent contradiction within Neoliberalism that sets the foundation for a much more blended approach. This contradiction is around 'intervention' and is described by Harvey (2005, p. 79) as:

> On the one hand the neoliberal state is expected to take a back seat and simply set the stage for market functions, but on the other it is supposed to be activist in creating a good business climate and to behave as a competitive entity in global politics.

As noted above in several places, such contradictions inherent in the discourse on Neoliberalism are many. As plainly stated by Harvey (2005, p. 152),

> The two economic engines that have powered the world through the global economic recession that set in after 2001 have been the United States and China. The irony is that both have been behaving like Keynesian states in a world supposedly governed by neoliberal rules. The US has resorted to massive deficit-financing of its militarism and its consumerism, while China has debt-financed with non-performing bank loans, massive infrastructural and fixed-capital investments.

Waeyenberge et al. (2011, p. 8) also note how "neo-liberalism has never been short of state intervention. Indeed, it has positively deployed it to promote not so much the amorphous market as the interests of private capital". Over time, this has shifted from the aggressive policy changes of the 1980s into, by the early 2000s, one of the many versions of a 'third way'.

As Troughton (2005) has argued for agriculture, the transformations away from Fordist production systems in the natural resource sector are incomplete and it is uncertain whether the theoretical delineations so carefully detailed within the literature actually appear in practice. Hayter (2000), for example, explores the circumstances of several new sawmill facilities in the BC forest industry. While job reductions (relative to productive capacity) have very much been achieved through the aggressive application of computer-managed production technology, there remain significant linkages to older production systems in terms of the organization of management / labour relations and the demarcation of jobs on the shop floor. For Hayter (2000, p. 269) the emergent system may be something of a hybrid "high-performance model and neo-Taylorist model" (see also Chapter 3).

This notion of hybrid is important for our forthcoming consideration and discussion of rural development moving into the 21st century. As fits with one of our opening observations about theory – we never expect to see such 'theoretical constructions' plainly acted out in the real world – there are always histories and cultures that transform the application. Hayter and Barnes (2012, p. 197) note that "when neoliberalism is grounded in particular places, it takes on hybrid forms, a result of local contingencies that are often found at those sites". They further argue that the inherent tensions and internal contradictions of Neoliberalism mark how "by its very constitution, neoliberalism cannot exist in the pure form of an unadulterated (unregulated) market. It must be joined to other social and geographic processes and relations" (Hayter and Barnes, 2012, p. 198). In rural development, while calls to Neoliberalize policy have been strident for 30 to 35 years, significant actors such as industry and the state have been reticent to let go of the old ways.

Disjunctures in policy behaviour, as we have seen in earlier discussions, became easily and comfortably accommodated within the ideology and practice of governments as they employ Neoliberal approaches to cover their withdrawal from responsibilities and their efforts to protect the taxpayer from budget deficits, but at the same time seem willing to re-deploy an almost 'neo-Keynesian' approach that commits re-regulation or massive taxpayer funds to bailing out economies / firms crippled by private-sector excesses.

With the breakdown of a Neoliberal policy approach, there has been an incoherent re-deployment of state support (usually debt and deficit derived support), and a reconfiguration of industry roles relative to communities and regions to deal with the contradictions – and real life implications – of decades of Neoliberal-inspired restructuring. The question is how might we characterize the current policy period for resource-dependent rural and small town places and regions? Coming up with such a characterization is challenging, because it needs to capture a number of divergent and / or conflicting trends or trajectories. Included among these are the many inherent contradictions of Neoliberalism itself, and the above-noted rush by numerous state governments (regardless of political affiliation) to deploy massive state spending and other interventions in the economy to support employment, production / consumption, and generally to provide stimulus during times of economic collapse. Also included are concerns about the 'end of life-cycle' status of much of the critical infrastructure in rural and resource-dependent regions (built during the immediate post-war decades, but run down by over two decades of Neoliberal policy) and the limited capacity of local governments to make the shift from a managerial to an entrepreneurial out-look (see Chapter 9) and from a governmental approach to a governance approach to community, regional, and economic development.

Senior governments are also being pushed by the activities and advocacy of rural communities and regions that are stressed by decades of neglect and a lack of re-investment. The state response has often been a retreat to older economic models (i.e. enhanced resource extraction, even where such activity no longer makes economic or policy sense) hoping to stimulate a short-term employment bump. Another state response has been a limited interest in var-ious forms of local or regional trust or royalty regimes. The notion here is to return to rural and resource-producing regions a very small proportion of the wealth that was generated within those regions and yet was consumed in dis-tant urban centres. Rarely are such schemes, however, of a scale and design necessary to support transformative rural change that can move resource-dependent regions into the 21st-century economy. All of these issues have raised questions about the most appropriate ways and mechanisms for re-investing in rural places, and about the appropriate roles for senior and local governments.

Added to the complexity is the changing role of industry. First shifting dramatically away from its historic paternal relationship with resource-dependent communities, industry is now having to rebuild those relationships under the guise of corporate social responsibility, impact–benefit agreements, and the like, as part of the community response to decades of neglect that is now demanding corporate initiatives to obtain a form of local 'social license' for new resource industry projects.

As noted, the blending of ongoing commitments to Neoliberalism, with a chaotic return to almost neo-Keynesian public investment policy, and a downloading of governance responsibilities to industry and communities, is creating a different policy model or framework. But the actions are too chaotic and idiosyncratic to each circumstance or jurisdiction to defy the kind of logic or coherence implied by the terms 'model' or 'framework'. Instead, we would label this new public and corporate policy period or era as 'reac-tionary'. This label incorporates a sense of the chaotic and disorganized ele-ments that support a level of policy incoherence. It also suggests a mix of retrenchment and opportunistic initiatives / reactions undertaken specifically to maintain a past hegemonic structure and set of actors / relationships.

In this reactionary era, hegemonic forces seeking desperately to maintain their positions of power (political, economic, etc.) are grasping at any form of initiative and activity that will secure for the short term their interests in a dynamically changing global and local social, demographic, political, eco-nomic, and environmental context. Ideological ties to Neoliberalism, for example, are readily sacrificed, abandoned, or combined with other approa-ches as needed. This loosening of philosophical or ideological commitment to policy approach in this more reactionary era allows actors to take up any position, at any particular point in time, so as to secure or advance their hegemonic position. The failure of major resource actors to transform dramatically away from older business models, the challenges state governments have had in 'getting off' their addiction to natural resource revenues and seriously support more value-added and alternative economic structures that make better use of the place-based competitive assets of rural regions, and the limited success of labour in identifying new pathways and tools for supporting workers and protecting worker rights in a competitive global environment increasing marked by mobile work and mobile labour, all reinforce the degree to which the historically hegemonic forces in resource-dependent regions have struggled with the accelerated pace of change in the global economy.

Closing

This chapter sets the stage for Part III of the book – a set of reviews of how major institutional 'actors' are responding to change in resource-dependent rural and small town regions. As such it provided background on concepts such as 'institutions' and 'institutionalism'. It then spent time exploring the transition from a post-war Keynesian political economy framework to a Neoliberal political economy framework. As noted, this transition has had a significant impact, on how resource development and resource-dependent community renewal have been approached. In the following chapters, attention now turns to those institutional actors, including large companies and small businesses, senior governments, communities and local governments, local service providers and the voluntary sector, and finally, individuals.

7 Large companies, big labour; small businesses, employees

Introduction

This chapter looks at one of the key institutional building blocks in resource-dependent rural and small town places – business and industry. This is an important topic, given the central role that a healthy economy plays in supporting healthy places. It is also important because the structure of business and industry in these small, resource-dependent places is a unique mix of the large and the small – all intimately bound together by a mutual dependence upon the strength of the local resource sector. Linked closely to these key economic actors are local and regional labour forces. Not only does their organizational structure mirror that of our dichotomist large and small categories, but labour is also intimately bound to the strength of the local resource sector.

To understand the changing political economy of resource-dependent places, we need to understand the basic context and contemporary strategies of business and industry. Therefore, this chapter is comprised of four sections. The first describes some of the structures, imperatives, pressures, and response strategies of large resource industry firms. The second looks at some of the key issues related to the organization of resource industry labour. In the post-war era, big industry led to the creation of big unions, and the restructuring of those industries has meant a significant change for the associated labour movement – both in terms of membership numbers and union organization. A final discussion looks at the opposite side of the economic structure in our resource-dependent rural and small town places – small businesses. As with the section on industrial firms, this section explores some of the general structures, imperatives, pressures, and strategic responses of small businesses.

Big companies

Industrial resource development is capital-intensive – and into the early part of the 21st century, it is becoming more so. In the post-Second World War period, the lesson of increasingly large industrial production, and the advent of communication and transportation technology that was beginning to connect global supply chains and industrial production so that it could be managed in real time, all play key parts in the drive for larger and larger resource-development industries. This section of the chapter looks at the general structures, imperatives, pressures, and strategic responses of large resource industry firms in the last 35 years of industrial restructuring.

Basic context

The basic operating context for large resource industry firms has been noted in various places thus far in the book. Their function in the global economy is that they extract raw natural resources, put those raw resources through a basic processing operation to concentrate them or to reduce their bulk for transportation, and then ship these minimally-processed resource commodities to value-adding higher-order manufacturing centres. Their products are typically low value-per-volume and so the economic model is to ship large volumes in order to recover capital and operating costs, and to generate a profit.

The size of the investment required to build the scale of operation capable of generating the needed volumes of resource commodity products has several implications. First, it continues to put pressure on the size of the firm and the need to grow larger in order to continue scaling-up production to counter the general trend of declining relative commodity prices over time. Second, there is a concomitant need to attract investment capital at an increasing scale to continue the growth imperative which also demands that large firms be able to satisfy the security conditions of lenders. Third, the scale and intensity of investment in low-value / high-volume commodity production leaves little room for product or economic diversification. This last point has been exacerbated by recent corporate strategies to deal with various crises of profitability.

This role and position in the global economy means that resource-production firms are very dependent upon global demands and prices for their commodity outputs. As the scale of production has increased, so has the level of vulnerability. As noted earlier in the book, globalization generally, and various types of international bi- and multi-lateral trade agreements in particular, have had the effect of opening markets to low-cost production regions. Coupled with the declining relative cost of ocean-going bulk and container transportation, resource commodities from low-cost regions are putting further downward pressure on prices and are displacing high-cost commodity suppliers in some of their traditional markets. As the global economy has accelerated, these pressures on large resource companies have challenged them into a constant state of readiness, adjustment, and response to both opportunities and threats.

Structures

The geographic organization of resource company operations is relatively straightforward. The basic processing or milling of the raw material occurs at facilities located in hinterland areas relatively close to the source of the natural resource itself. This is because the basic processing task aims to reduce the bulk of the raw material into a commodity suited for export. The cost of transportation relative to other inputs such as energy and labour determines the location of these basic processing or milling facilities. The head offices for resource companies, however, do not tend to be located in hinterland areas. Instead, they are generally located in metropolitan or core areas close to the political and financial decision-making centres relevant to their business and also where they can easily access high-order business services including finance, insurance, legal, etc. Depending upon the regulations in place within individual states, multi-national resource companies may maintain a subsidiary branch office in countries where they operate, or they may manage activities direct from a foreign office.

But who are these large resource companies? In general terms, business analysts would organize them into three basic types – although we recognize that there are lots of hybrid forms as well. A first general type would be the 'private company'. In this case, the company would be owned by an individual or by a small group of individuals. The private firm has a number of advantages in that the governance structure is relatively simple, the owners and (usually) the senior managers tend to have a lot of knowledge and history with the company and the industry, and planning for profitability and company survival in both the long and the short term is dictated rather simply by the wishes and fiscal capacity of the owner / ownership group.

Private firms also have a number of challenges relative to other ownership types. Among these challenges are difficulties in accessing large amounts of capital such as publicly traded firms can do via a share offering. Another is that management boards of directors can provide other types of companies with access to a wide range of skills and information, as well as to networks of influence, that private ownership cannot. As the imperatives of size have worked their way through many resource industry sectors, we find fewer privately owned firms as they either are bought up by larger publicly traded companies, or they convert themselves into publicly traded companies.

A second company type, and another type that is becoming less common in developed economies, is the state-owned company. These state-owned companies can develop from the ground up when the state sees an opportunity to capture wealth from a resource industry. This was certainly the case in Norway, when the state was confronted with a number of options when it was poised to exploit North Sea oil and gas development opportunities. One option – the one used in most developed economies under an increasingly Neoliberal policy framework – would have been to simply sell resource access rights to a large multi-national corporation and then realize returns in the form of royalty or taxation benefits based on the volume of commodity extracted. Another option – and the one Norway eventually adopted – was to get into the industry itself with a state-owned company so as to realize not just royalty or taxation benefits, but also corporate profits. In Norway, the state owned the company Statoil ASA is its oil and gas development vehicle.

What began as a public company has also transitioned into a publicly traded company – although to protect its policy goals, the Government of Norway maintains about two-thirds of the shares. Norway was able to invest its North Sea windfall profits in a legacy fund that today often generates interest that exceeds the size of the national state budget.

In news coverage of resource-development debate in Canada, the gap between the wealth being generated by resource development and the revenues available to the state to meet social and economic development needs, and the ability of the state to develop or build a legacy fund from both non-renewable and renewable resource extraction has been highlighted (Simpson, 2012). Tencer (2014), for example, compared the resource legacies of two 'oil economies' (Norway and Alberta) and reported that while Alberta is carrying more than CAN$7 billion in debt, Norway has a resource trust fund that exceeds CAN$905 billion.

Many states have had such state-owned enterprises. In some cases, this was a matter of national policy in the Keynesian post-Second World War era. In other cases, it was a necessary step, as industrial capital was either too slow or unwilling to invest. In still other cases, it provided a stop-gap measure as the state stepped in to purchase and then operate a failing resource industry when industrial capital sought to withdraw.

Connelly and Nel (2017a) identify how state enterprises were a significant feature of development policy in New Zealand, playing roles in the development of both regional and national economies. In Finland, Tykkyläinen (2006) also highlight the role of state-run enterprises and the transitions that such enterprises went through as Finnish public policy changed over time. In most cases, the closure or privatization of these industries in the economic and policy transitions of the Neoliberal era had significant negative impacts on local and regional economies and employment. These negative impacts were exacerbated by the pace at which the changeover occurred and by the lack of attention to transition planning. While many states have experience with state-run enterprises in various natural resource sectors, and, despite the very successful example of Norway's state engagement with the oil and gas sector, current Neoliberal public policy approaches dissuade this type of government investment in both businesses and the market more generally.

The third, and most common, corporate structure in the various resource industry sectors involve publicly traded companies. While these publicly traded companies do vary in size, the long-term pressures around scaling-up production to be price-competitive has meant that generally these companies are getting larger and are becoming more multi-national in their activities. While neither of these trends is particularly new, they do seem to be accelerating as processes of globalization unfold. Among the consequences is the more recent transition to increasingly oligopolistic or monopolistic resource economies.

As noted earlier, publicly traded companies have some advantages over the other two general forms of businesses. First among these is access to capital through stock offerings to shareholders. This has the potential to generate significant sums at times when companies may be looking to undertake extensive new expenditures or develop new projects. This benefit comes with a caveat in that shareholders expect to receive a return or dividend for their investments, and this 'payback' expectation puts pressure on the companies to perform financially on a regular basis.

Another advantage for publicly traded companies is that they are required to have a diverse board of directors. In organizational theory, such boards not only have required duties of governance and oversight, but they also provide the company access to external information and networks of support and influence that can be of assistance. In many cases, boards of directors can be the focal point for the development of strategic linkages and alliances for companies.

However, one of the significant challenges for publicly traded companies is related to their first advantage – investment by shareholders. Both law and practice require that companies be attentive to shareholder investments and mindful to the returns to shareholders that are generated on that investment. Quite often, especially in circumstances where a company may be struggling financially, demands for shareholder returns can lead to short-term decision making aimed at generating a profit that in the long term works against the company's sustainability. At a general level, the contradictions of short- versus long-term thinking and strategizing on the part of capital is clear. As noted by Porter and Kramer (2011, p. 64) in the Harvard Business Review,

> A big part of the problem lies with companies themselves, which remain trapped in an outdated approach to value creation that has emerged over the past few decades. They continue to view value creation narrowly, optimizing short-term financial performance in a bubble while missing the most important customer needs and ignoring the broader influences that determine their longer-term success. How else could companies overlook the well-being of their customers, the depletion of natural resources vital to their businesses, the viability of key suppliers, or the economic distress of the communities in which they produce and sell? How else could companies think that simply shifting activities to locations with even lower wages was a sustainable 'solution' to competitive challenges?

Trends

As noted earlier, the general cost-price squeeze of natural resource commodities, the entrance of competitors from low-cost regions of the world, and the acceleration of demand boom-and-bust cycles for resource commodities have all put pressure on resource industries in developed economies. In this section, we explore three forms of strategic response by industry: cost minimization, risk sharing, and global reinvestment.

Cost minimization

Crises of profitability, and the challenge of competing for market share that result from the cost-price pressures listed above, all support strategic moves by resource companies to minimize their costs of production. Whatever approach is taken, the goal is to reduce the price per unit of commodity as delivered to customers.

One of the first of these cost-minimization strategies involves an aggressive substitution of capital for labour. While such has been found in most manufacturing sectors, largely as a result of the rapid development of computer-controlled robotic manufacturing equipment, it is an acceleration of older trends. But as pressures around competitiveness and profitability rose, the drive to minimize labour costs occurred across natural resource sectors. Recent studies in New Zealand (Connelly and Nel, 2017b), Canada (Halseth et al., 2017b), Australia (Argent, 2017b), and Finland (Halonen et al., 2017) all highlight dramatic employment losses that continued through the 2000s (see also Chapter 5).

In concert with the reductions in labour force numbers has been a long-term effort at restructuring the terms of the natural resource workplace as well. In Australia, Argent (2017a, pp. 145–146) notes how:

> As with many other long-standing social and economic institutions built up over the Keynesian-Fordist era of the post-Second World War boom, Australia's centralised industrial relations system underwent ongoing and substantial 'reform' during the 1980s, 1990s, and 2000s. Consistent with the nation's drive for global free trade, consecutive federal and state governments over this era, regardless of political stripe, embraced an agenda of macro-economic reform that would putatively raise the competitiveness and productivity of every branch of industry and government … In the context of these institutional changes, amidst other far-reaching changes in Australian society and economy, union representation has gradually but consistently declined. Between 1990 and 2009, levels of union representation in Australian workplaces approximately halved from just over 40% to 20%.

The same policy changes to labour regulation are found in other developed economies – an outcome of the Neoliberal policy turn and the recurrent crises of profitability within natural resource sectors. Reducing the costs of production by reducing the number, and cost, of employees has been a key part of the industrial strategic response to economic restructuring and globalization.

Another key part of reducing production costs for resource-based industries operating in developed economies is to seek reductions in their regulatory burden. As part of the Neoliberal policy discourse on reducing government involvement in the economy, industry has engaged in an ongoing dialogue with governments to reduce or remove rules, regulations, and activities that have accumulated in the policy environment. One particular, and more recent, area of policy change has been the various types of environmental assessment processes or regulatory approval processes for proposed new resource-development projects. Over time, and in response to crises, gaps, or needs, multiple sets of such approval or review processes may have been set up within different jurisdictions. It certainly does create an uneven playing field in terms of the approval processes that projects may encounter between different jurisdictions.

In these cases, industry advocates have called for the streamlining of approval processes. In many jurisdictions, such has been the case. In their review of environmental assessment processes from the lens of understanding the cumulative impacts of multiple projects developed over different time frames and across different spatial scales, Halseth et al. (2016, p. 7) make note of the

> repeated calls in recent years for more streamlined processes by some governments and project proponents. The saying that *time is money* can seem very much the case when resource development projects are linked to short-term windows of opportunity in global markets.
>
> (italics in original)

In addition to labour and regulation, one of the cost areas aggressively targeted by resource industries has been the various forms of taxes or royalties that they pay to different levels of government. Depending on the institutional structure of different states, there may be one, two, three, or sometimes even four levels of government to which some form of taxation is paid. Not only is this a complex matter to track, and even more complex to predict what tax impacts will be on a company's operating bottom line in coming years, but companies often feel that they are paying multiple times for the basic services that they receive back from these levels of government.

A particular focal point has been taxes or costs at the local level. In one community in northern BC, a community meeting was held with the managers of a local sawmill. The mill had been idled for a number of years and the dialogue was in advance of an expected start-up of operations. When asked about the fiscal health of the proposed start-up, a corporate executive stood up to answer. So entrenched had become the anti-tax mindset of industry that the executive simply answered that they could make the mill run but that local taxes were killing them and that everyone needed to convince the town government to reduce the tax burden. The executive's comments were greeted with stunned silence. Finally, an audience member stood up and asked if the executive knew what the local government tax bill was for the mill – to which the executive replied that he did not know the specific number, but that it was just too high. Another audience member then stood up and said that the mill's tax burden had long ago been reduced and capped at a sum less than that of a typical single-family residence in the community despite the fact that the value of the mill was more than 100 times the value of any home in the community. The incessant mantra of Neoliberal arguments can roll forward without evidence or without cause.

Another local strategy is for companies to seek temporary relief from local tax expenses, especially during times of economic downturn or when profitability is under threat. Halseth et al. (2017a) highlight one such occurrence. In the years following the 2008 global economic collapse, each of the three major forest products firms operating in Mackenzie, British Columbia, were granted two-year municipal tax exemptions. Mackenzie Pulp received exemption for the tax years of 2011 and 2012, Canfor for the tax years 2010 and 2011, and Conifex for 2011 and 2012. The exemptions:

> were designed to assist with restarting the mills and getting the local operations of the company into a more competitive economic position. The exemptions were dependent upon the companies beginning full-time operation, with stipulations generally including minimum levels of employment, as well as average annual and weekly operating schedules.
>
> (Halseth et al., 2017a, p. 284)

These are difficult issues for all involved. The companies are under significant economic and shareholder pressures, and so are searching for all ways (on both the revenue and expense side of the ledger) to grow profits. Communities are dependent upon the jobs these industries provide and are eager to keep the mills / mines / etc. open as unemployed workers are just as likely to punish local governments at the electoral ballot box as they are to criticize companies or their own unions. However, as many communities also rely upon industrial taxes for a significant part of their operating revenues, to grant such tax concessions places those same local governments in a fiscally vulnerable situation unless they cut services or otherwise transfer the tax burden to local businesses and residences.

In writing about the pulp and paper town of Espanola, in northern Ontario, Keenan (2015) noted that for decades the large resource industry mill had

> been the biggest contributor to the town's tax revenue. An appeal board found the drastic slump in the forest industry meant the value of the mill had declined. So, Espanola was forced to refund $5.2-million to Domtar for the municipal taxes it paid between 2009 and 2012 …
>
> From Kitimat, B.C., to Halifax, and in almost every municipality in Ontario, some of Canada's biggest companies are winning reductions in the values tax assessors have placed on their properties, often because their factories and mills have cut production. Decisions in favour of the companies mean some towns and cities have been forced to emulate Espanola and issue refunds – in some cases amounting to millions of dollars.

The loss of tax revenues from pulp-and-paper giants, auto makers, steel producers, retailers and other companies has caused a massive shift in the burden to residential taxpayers, staff layoffs in some municipalities, and delays or cancellations of upgrades to roads and other infrastructure.

A final corporate strategy tied to revenue / expense balance sheets has to do with changes in maintenance and upkeep of infrastructure and equipment. Various strategies have been used in this regard. A first has to do with strategically ranking maintenance needs, via some form of rating system. For example, green lighting a maintenance issue means that there is no problem for the operation of the facility or the safety of the workplace. Yellow lighting a maintenance issue identifies that, while it should be fixed, it is not currently impeding production and does not pose an imminent workplace safety risk. Red lighting a maintenance issue identifies that this issue must be fixed, as it directly impacts either the production process or workplace safety. Another is deferred maintenance, especially within jurisdictions where capital investments can be written off against taxes. Deferred maintenance is something that can be done at times of high profit margins, so as to reduce the level of taxes that the company pays to government.

Whatever the corporate strategy around infrastructure and equipment maintenance and replacement, some companies will use a deferred maintenance strategy as part of a long-term exit plan from an industry or a region. It makes no sense, after all, to fix everything up in a plant that you are planning to close anyways. As Markey et al. (2012, p. 30) note for BC-based resource companies now investing heavily in low cost regions: "Investments in the southern United States and outside of North America underscore the late stages of industrial models that would run down existing assets while repositioning to be more competitive in lower cost commodity-producing regions".

Finally, beyond labour and the pressure to report and return profits to shareholders, the strategic moves across resource sector companies have also weakened the strategic planning capacity of companies within those sectors. In this case, we are talking about the hollowing-out of companies by their reduction of the middle management class. These operational costs were as aggressively reduced as any others in the restructuring process as part of cost-minimization efforts. What was left was the core business activities of shop-floor production and senior management administration.

The impacts of this hollowing-out of resource industry companies has been significant as latter stages of a global economic restructuring, together with climate change, and renewed political tension in the world has unfolded. Companies now lack the capacity and talent to undertake strategic planning within this dynamic and rapidly changing context. They lack the capacity and staff to undertake visioning and option planning to help guide senior management decision making. This has left firms in a weakened competitive position.

Risk sharing

Another strategic response by corporate capital to the uncertainties of the global resource marketplace, and to the need for significant capital in order to undertake very large resource-development projects, has been the sharing of risk, largely via joint ventures, and the off-loading of risk mainly via sub-contracting. The first strategy is an old and well-established mechanism with respect to corporate investments and decision making in natural resource sectors.

In the 1970s, industrial resource investment in BC's forest sector combined the financial and technical expertise of multi-national corporations with the access to a local forest resource base held by BC-based firms in order to support a significant burst of investment in the pulp and paper sector within the interior part of the province. In that period, three new pulp mills were built in Prince George. The Prince George Pulp mill was a joint venture by Canadian Forest Products and the Reed Paper company of England. The Intercontinental Pulp and Paper mill was a joint venture by Canadian Forest Products, Reed Paper, and Feldmuehle Ag from Germany. The Northwood Pulp mill was a joint venture by Noranda Mines and the National Forest Products group (comprising local companies Sinclair Spruce Mills and Upper Fraser Lumber Mill). In 1981, the Cariboo Pulp & Paper mill was built in Quesnel. The mill was a joint venture of West Fraser and Daishowa-Marubeni International Ltd. (DMI). DMI is an integrated forest products company established in Canada in 1969 by Daishowa Paper Manufacturing of Japan and the Marubeni Corporation of Japan.

In the early 1980s, during a significant global recession, an otherwise Neoliberal-sympathetic BC government sought to reinvent a Keynesian public policy approach by kick-starting economic activity and initiating a coal mining boom in the northeast section of the province. Built around a set of long-term sales agreements with a number of steel mills in Japan, the project was massive in scope. It involved a new port facility in Prince Rupert, upgrading and extending rail lines from the port to the new mines, extending road and power line networks to the new mines, and construct-ing an entirely new town site in the wilderness to house the mining work-force. Two very large open pit coal mines (Quintette and Bullmoose) were developed. When the Quintette mine started in the early 1980s, the own-ership structure included Denison Mines (49.95%), Tokyo Boeki (20.10%), and Mitsui Mining (20.10%). The remaining shares were owned by a number of smaller firms, including two French companies. When the Bullmoose mine started in the early 1980s, the ownership was divided between Teck at 51%, Nissho-Iwai at 10%, and Lornex at 39%. Lornex, in turn was held 21% by Teck and 68% by Rio-Algom. In 2000, just a little over a year from its final closure, the Bullmoose mine ownership was divided between Teck (61%), Rio Algom Ltd. (29%), and Nissho-Iwai Ltd. (10%) (Halseth and Sullivan, 2002).

In these joint ventures, at least four motivations can be identified for corporate participation. First, obviously, is that the scale of the investment and the risky nature of resource commodities underwrite decisions to share risk in the gathering of needed capital. Second, joint ventures can link firms with different skill sets, knowledge areas, and experience which may help make the overall project a success. Third, and related to the second, is that different companies in the consortium bring different corporate assets – such as a secured access to a resource, an already recognized brand in the marketplace, or a roster of existing markets and clients. Fourth, companies may partner in joint ventures as a way to enter a sector where they currently have a limited profile, limited holdings, or limited experience. In this case, joint ventures may be a relatively quick way for a company to establish a presence in that sector.

More recently has been that during the mid-2010s, there has been a rapid increase in interest around liquefied natural gas (LNG). A number of factors supported this interest, including the desire by Asian economies to change their energy profile (Japan and South Korea seeking to replace nuclear; China seeking to replace coal) and that regions of Canada, the United States, and Australia primarily were reporting massive new natural gas reserves if accessed by hydrologic fracturing (fracking) technology. This led to a LNG 'rush' with massive speculation and projects in the three source states identified above, as well as numbers of other places.

In Western Australia, the (LNG) industry is much further developed than in some of the other potentially competing jurisdictions. In 2016, the Government of Western Australia's Department of State Development reported that their state was already home to three operating LNG export plants and they were expecting another two to come online over the next two years. The existing operations include the North West Shelf (16.7 million tonnes annual capacity), Pluto (16.7 million tonnes annual capacity), and Gorgon (15.6 million tonnes annual capacity) projects. The North West Shelf LNG project involves a consortium of six equal partners. These include: BHP Billiton Petroleum (North West Shelf) Ltd.; BP Developments Australia Ltd.; Chevron Australia Ltd.; Japan Australia LNG (MIMI) Ltd.; Shell Development (Australia) Ltd.; and Woodside Energy Ltd. Also included in the partnership, though not involved in the plant infrastructure, is the China National Offshore Oil Corporation (NWS, 2016). The Pluto Project, in contrast, is operated by the Australian-based and publicly traded company Woodside Energy Ltd., whose parent company is Woodside Petroleum Ltd. (Woodside Energy Ltd., 2016). For the Gorgon Project, as of 2016, the ownership structure of this joint venture included the Australian subsidiaries of Chevron (47.3%), ExxonMobil (25%), Shell (25%), Osaka Gas (1.25%), Tokyo Gas (1%), and JERA (itself a joint venture of Tokyo Electric Power and Chubu Electric Power) (0.417%) (Chevron Australia, 2016).

Sub-contracting

A further strategy around risk sharing has been for companies to off-load activities that used to be done within the companies themselves. This exercise of sub-contracting has numbers of advantages. To start, sub-contracting off-loads the ongoing wage bill, together with the payroll and pension expenses, that comes with employees. It also means that instead of full-time employees, one only needs to hire a sub-contractor when that activity or service is specifically needed. Further, employees are an expense against profitability, while the costs of sub-contractors can sometimes be written off against the corporate tax bill.

Depending on the operating circumstances of companies, sub-contracting operational activities can also transfer risk and defer costs. Consider a typical sawmill in Canada. Historically, the transportation of logs from the forest to the sawmill was done via company employees driving company trucks. Today, such work is almost universally done via sub-contractors – small companies or independent owner-operators. The truckers now have to buy and maintain their own trucks and compete with other truckers for log-hauling contracts. Companies often put 60- or 90-day payment clauses into such contracts, meaning that truckers are out-of-pocket for expenses until paid. It also means that if something catastrophic happens, such as a company declaration of bankruptcy or closure, these independent truckers are not going to get paid at all, as they are typically far down the list of creditors, which is topped by shareholders, banks and large lending institutions, etc.

A further aspect of sub-contracting that is worth noting is the flexibility given to companies with regard to monthly operating costs and revenues. If a company is the 'only game in town', they can put pressure on sub-contractors to accept lower than contracted fees when profits are at risk. The sub-contractors have little choice, as they are so intimately tied to the company for their survival that, like the earlier story of local governments, they will usually agree to the company's demands.

Globalizing investments

A third strategic response by resource industry companies to the sweeping changes, pressures, and opportunities of globalization has been to actively invest in alternative production locations. In many cases, this would involve investments specifically in low-cost production regions in order to take advantage of the cost and regulatory advantages for accessing natural resources while still maintaining their corporate strategic ties with, and linkages to, existing markets.

Using the profits generated by resource production in the rural and small town regions of developed economies in order to fund investments in low-cost producing regions is not a new strategy or phenomenon. Bradbury and St. Martin (1983) told the story of the Iron Ore Company of Canada's (IOC) global initiative as the backdrop to their work in the town of Schefferville, Quebec. Developed as part of large iron ore mining region, at the time of their research the mining activity in Schefferville was winding down and the town was losing people and economic activity. As noted by Bradbury and St. Martin (1983, pp. 138–139),

> A further contributor to instability in regional communities during a winding-down period is the policy of 'second-sourcing' by the company. This is a procedure whereby a company acquires a second (or more) source of raw material or establishes a parallel production line to compete better in the world market. It means, however, that companies have the opportunity to close and open mines depending upon the price of labour, on the dominant labour conditions, on technological changes, and on the declining quality of an ore body ... In 1964, the American-based Hanna Mining Company, part-owner of IOC, opened a mining complex in Brazil (the MBR Aquar Claras project). While there was no direct ore substitution between IOC and Hanna Mining in the Quebec-Labrador region, the tendency for large companies to develop alternative sites and sources of raw materials has been a common feature of the period since the Second World War. The low cost of labour in underdeveloped countries has led to increased competition between suppliers from developed and underdeveloped countries.

The iron ore mines in the Schefferville region were subsequently closed.

There are, however, challenges for to the mobility of large resource companies. In respect of the global mobility of capital, Hayter (2000, pp. 328–329) notes that large multi-national corporations

> have deep roots in particular places as a result of their access to particular resources, huge capital investments, and the accumulated know-how of workforces ... It has not been easy for already invested forest-product capital in BC to simply run away. New capital spending, of course, always has alternatives.

In light of those alternatives, Markey et al. (2012, p. 30) recently have traced the story of investments by leading BC forest products companies in the southeastern parts of the United States. For West Fraser, they note:

> In their 2007 annual report to shareholders, West Fraser Timber (2007: 5) reported that they had 'completed the acquisition of 13 sawmills in the southern United States [Texas, Alabama, Florida, Georgia, North Carolina, and South Carolina], adding to our two existing mills in Louisiana and Arkansas. West Fraser is now a truly North American wood products company. We are one of North America's largest lumber producers with significant panel board and pulp and paper assets as well.

Approximately one-third of our lumber production is now based in the United States. This is an important strategic step for West Fraser as it significantly increases our geographic, product and currency diversification.'

Big labour

In developed economies, the rise in the post-Second World War period of very large resource industries, owned and managed by very large, often multi-national, companies, was accompanied by the need for labour and the rise of very large labour organizations. This part of the chapter explores some of the reasons behind the rise of trade unions in resource industries and the consequences for organized labour of dramatic industrial restructuring after 1980.

Background

To start, there are several reasons suggested in the literature as supporting the expansion of unions in the long boom period of resource development in the rural and small town regions of developed economies. Some of these are practical, and some ideological – but all fit with the political economy context of the times and the places.

On the practical side, the expansion of Fordist-style industrial develop-ment was ideally suited to unionization. Very large workforces, organized into very discrete tasks and roles in the production process, required multi-ple layers of management. At one level, capital needed a framework within which workers could be organized, the day-to-day management of produc-tion could be controlled, and a clear set of rules which the entire system could know and learn, and by which the entire system could work. At the same time, workers needed to know the rules in these new, massive, pro-duction settings. They needed to know that there would be a certain stan-dard of fairness to treatment and pay, that there would be due process should problems arise, and that order rather than chaos would now govern the shop floor. For both capital and labour, the new environment of massive industrial resource developments required the discipline that labour contracts under a union-organized shop would bring.

But there were also historical and ideological motivations behind the evo-lution of big labour. In the Cold War era, many Western democracies were deeply concerned about the 'happiness' of the labouring classes. The relative stability that unionization could bring was one of the bulwarks against com-munism, whereby unions, and by extension, workers, could be bound into the capitalist project through their benefits-sharing (wages and benefits) arrangement with companies.

Coincident with these favourable trends was also the election in numbers of countries of labour-oriented governments. Not only did such governance continue Keynesian-style public policy approaches, but many actively enshrined rights and other obligations around labour organizing into state legislation. As discussed by Connelly and Nel (2017b, p. 221),

> Late nineteenth and early twentieth century New Zealand was characterized by profound social changes, not least among these was the entrenchment of a strong culture of unions, labor protection, and social welfare. Mining and harbor unions led the way in successfully pioneering the realization of employment and industrial agreements (Maunder, 2012). In parallel with changes in social welfare provision, this led to the creation of one of the Organisation of Economic Co-operation and Development's (OECD's) most egalitarian states by the post-World War Two era (Stanford in Cooke et al., 2014). In parallel, by the 1970s the country had one of the most controlled economies in the OECD.

Together, these practical and ideological foundations fit the organization of labour in various resource sectors into a wider project supporting expansion of the social welfare state and expansion of industrial resource development. While capital paid wages and taxes to support many aspects of the welfare state, so did these now organized workers themselves pay taxes to support the same social welfare state activities.

Both capital and labour benefited from the expansion of the social welfare state in the immediate post-Second World War period. Health care and education assisted with the reproduction of labour and the establishment of unemployment support schemes also helped to maintain a surplus labour market when not directly needed by capital. Pensions helped to join the state, capital, and workers in the provision of a public good.

The implications of the organization of labour under a Fordist production model and under union organization were significant. As discussed in Chapter 3, labour contracts set in place very detailed job descriptions and clear job hierarchies. This also fit well with a Keynesian policy approach that favoured high levels of employment.

Another implication was the extension over time of increases in wages and benefits. In a low-value, high-volume, product environment, capital had an imperative to keep production going in both good and challenging market conditions, but when market conditions were good, it was especially vital to keep production going. This gave labour an advantage in bargaining and over time, union labour in most resource sectors became very highly paid relative to other parts of the economy.

A third key implication of a unionized work environment was the application of rules around seniority and training / qualifications. As part of a system designed to promote merit and to protect against capricious management, these rules became well entrenched in both labour contracts and the labour movement more generally. The consequences of these rules would have serious implications for communities and capital in the post-1980s era. Finally, the organization of labour established a distinctly uneven geography of employment in resource-dependent places. On the one hand, there were the workers in the resource sector. They typically had high wages, an extensive benefits package, unionized job protection, and their work was generally regular and full-time. In contrast, the remaining workers in the service sector of these same resource-dependent places typically had much lower wages, few benefits, no union protections, and their work was often part-time or casual / on-call. The presence of high industry wages and lower service sector and, in some cases, professional wages has created 'two-speed' economies, producing a socio-economic divide between different residents who are able to directly benefit from industry employment (Plummer and Tonts, 2013; Wetzstein, 2011).

Transition

However, the transitions that began to impact the organization and workings of capital during the late 1970s and beyond also had a significant impact on labour. Processes involving accelerated globalization of the economy, the production transformation from Fordism to Flexible accumulation, and the public policy transition from Keynesian to Neoliberal policy approaches all acted to reduce the relative size of the resource sector workforce and weaken the role of organized labour in numbers of industries.

To start, the increasing globalization of the economy had tightened the competitive environment for companies. Production in high-cost regions could sometimes not compete with new industrial entrants. Depending on the resource sector, and the state, the post-1980 period has been marked by the closure of many resource production sites and loss of those well-paid jobs.

As recounted earlier, the transformation of production from Fordist to more Flexible models also had a significant impact on labour. Massive job losses reduced both the viability and the bargaining power of unions. The subcontracting of work previously done by workers also acted to reduce the bargaining power of unions and created worker-versus-worker conflicts during times of labour strife. Among the impacts of this reduced size and bargaining power was that unions sought increasingly to amalgamate with one another in order to maintain their economies of scale (Halseth et al., 2017b).

A third significant arena for change involved the Keynesian to Neoliberal public policy transition. In one of the most dramatic instances of public policy shift, Connelly and Nel (2017b) describe New Zealand's shift from one of the most progressive social democratic states to a very open Neoliberal state. Drawing upon a mining region example, they describe the dramatic impacts upon the role of unions in supporting workers and communities:

> Within this broad context, the once powerful unions, which had their genesis in these regions and which had been advocates of worker rights and community well-being in the face of exploitation since the late 1800s,

lost their voice as the state, in support of business, introduced more flexible employment conditions, which over time has seen the casualization of labor, contract-based employment, and the fly-in fly-out system on the remaining mines. This weakened the once powerful bond of solidarity between unionised workers and small town communities, further depriving local communities of a voice and local organizational skills, and weakening their ability to respond effectively to economic change.

(Connelly and Nel, 2017b, p. 318)

These types of policy changes, and their impacts, were widespread across OECD states. In Australia, for example, Argent (2017a, pp. 145–146) describes the shift from a 'workers' paradise' to one with Neoliberal labour regulation:

Australia's centralized industrial relations system underwent ongoing and substantial 'reform' during the 1980s, 1990s, and 2000s. Consistent with the nation's drive for global free trade, consecutive federal and state governments over this era, regardless of political stripe, embraced an agenda of macroeconomic reform that would putatively raise the competitiveness and productivity of every branch of industry and government. New institutions, such as the Productivity Commission and National Competition Council were formed to drive this agenda and ensure adherence to it by public and private sector players alike. In the arena of industrial relations, the old corporatist arrangements between state, capital, and labor were maintained under the Hawke-Keating Federal Labor Governments (1983–1996) but the various 'Accords' negotiated between the parties concerning wages and conditions also contained the need for ongoing increases in worker productivity as a trade-off for continued improvements in the welfare safety net and wages growth. A key element of this productivity drive that emerged over the various Accords was 'enterprise bargaining' – the negotiation of wages and conditions between business owners and workers at the scale of the individual business. Theoretically, at least, this approach was a boon for employers for it would allow for the better matching of wage levels and worker productivity than that provided by the hitherto centralized arbitration and conciliation process (Dabscheck 1994). The conservative Howard-led Coalition Federal Government (1996–2007) attempted more radical reform of the industrial relations system, seeking to further remove the influence of labor unions in workplace representation while encouraging greater flexibility in bargaining arrangements (Fagan and Webber 1999; Ellem 2006). In the context of these institutional changes, amidst other far-reaching changes in Australian society and economy, union representation has gradually but consistently declined. Between 1990 and 2009, levels of union representation in Australian workplaces approximately halved from just over 40 percent to 20 percent.

While each of these pressures of globalization, changes in production, and changes in policy have been contested, the long-term result has been definitive. Resource industries in developed economies now operate with far fewer workers than in the past, are more prone to temporary closures as multinational firms balance production costs across their numerous operating regions, and are more exposed to the accelerated pace of economic booms and busts in the contemporary global economy.

Small business / workers

The latter part of the chapter turns its attention to the other side of the economy in resource-dependent rural and small town places and regions – small businesses and the employees who work in them. This part of the chapter begins by setting a context for small business and then explores some of the implications of economic restructuring on those businesses and their workers.

Context

Rural and small town places are very challenging settings for small businesses. There is generally a small market and local businesses tend, therefore, to be generalists rather than specialists – a general store instead of an electronics store, for example. Because of the small market, businesses also have few opportunities for economies of scale and this very often results in higher prices for goods that are bought locally. The lack of local competition can also account for a lack of price competitiveness. Depending upon how close or far the nearest large urban centre, residents may have limited options for accessing other stores or service suppliers.

Most rural and small town places have a basic array of services that marketing and economic geography link to both the size of the community and the size of its trading or market area. Typically, small service sectors cover such areas as food, hardware, drug stores, health care services, local government services, and numbers of other small shopping opportunities – again, dependent upon the size of the community and the size of its trading / market area.

Over time, different external pressures have impacted this local business environment. One of the first of these pressures has been improvements to the transportation network. On the one hand, improvements may mean that local shoppers can now easily commute to distant large centres for shopping, thus exacerbating retail sector leakage at the local level. On the other hand, area residents who may have come to the small community to shop may now simply by-pass local stores in favour of those distant – but now closer in terms of travel time – shopping centres.

Small businesses in rural and small town places often tend to be owner-operator businesses. Many of these are the sub-contracting companies closely engaged with the local resource industry referred to earlier in the chapter. Others might include small restaurants, shops, or retail stores. The owner is often the principal employee. Where other employees are required, there is often a ready supply in resource-dependent communities. This ready supply of labour often comes in the form of the partner / spouse of a person working in the resource sector or young people entering the workforce for the first time. In the case of partners / spouses working in these small business and service sectors, the gender division of labour that has persisted in resource industries means more often than not that these are women. As noted earlier, these jobs tend to be part-time, lower-paid, with relatively few benefits, and with limited labour organizations such as unions. While not always the case, and recognizing that historic gender divisions of labour are eroding in all sectors (Reed, 2003b), this characterization is still quite common (Status of Women Canada, 2013).

The small business sector in resource-dependent rural and small town places has also been marked by a limited number of chain or 'brand associated' businesses. In banking, automobile gas and service stations, food stores, insurance, and real estate, chain business franchises have always been common. Over time, many national and international chains (especially in the hyper-competitive world of fast food retail) have begun developing business models that are better suited to the smaller markets of rural and small town places. The rest of the small business community tends to be made up of independent businesses.

Dependency

As noted in the previous discussion on big labour, most of the economy in rural and small town places is driven by the large numbers of well-paid jobs in the resource sector. Small businesses in these places are, therefore, very dependent upon the health of the local resource industry for their own economic survival. To start, many of these businesses may be part of the service or supply network directly tied to the resource industry. They may be sub-contractors or parts and service outlets that make most, and sometimes all, of their income from direct work with the resource sector.

Even small businesses that do not directly supply goods or services to the resource industry are also very dependent upon the health of that industry for their own survival. They depend upon the wages earned in that industry to support household purchases. As in any model of a local economy, the wages are an input to the local economy earned via the production of resource commodities that are exported. That basic sector income is then circulated and re-circulated through local business and local spending to support the local economy. The denser the local network of businesses, the more likely the local economy is to capture more of the overall economic benefit.

This dependency upon the local resource sector has meant that the level of risk for small businesses is also high whenever disruption to the dominant resource sector is encountered. Not only do local businesses act as a booster for the community and its small business sector, but they also tend to act as a booster for the resource sector as well. Changes that could negatively impact that resource sector in terms of job losses or potential closures represent a direct and immediate economic threat to the small business sector. As a result, these local economy actors can serve as a dampening mechanism on discussions of change as they may advocate for the maintenance of the established economy.

As with large businesses, small businesses also tend to be risk-averse. As a small business person, they are already risking a great deal in establishing and operating their business. Economic transformation is difficult in this setting. The political economy of small business tends to work against innovation and change. That noted, we also add that this is not always the case, as some small businesses turn their concerns about risk and dependence into a motivation for innovation and change as part of a truly entrepreneurial small business survival strategy.

In terms of being a force for change and resilience, or for supporting the established economy, small businesses in resource regions have a limited infrastructure for helping to speak for their interests. Chambers of commerce may exist, but due to their small size, their effectiveness and operation may fluctuate over time. A good deal of the influence that small businesses wield in the community comes from the personalities of the small business owners themselves. They tend to be leaders in the community, and they may take either informal or formal roles in that regard. Because small businesses typically depend upon interpersonal relationships and knowledge for their success, these tend to be cultivated and they can form a circle of influence in the small community.

Transition

In the immediate post-Second World War period, the small business environment in resource-dependent places often tended to thrive. Population growth, expanding employment opportunities, the baby boom, and weak regional transportation systems all supported local shopping. Compared to today, the retail sector in those places had a wider array of goods and business types than are currently seen.

As the late 1970s and early 1980s began with both recession and then an extended period of economic restructuring and transition in the resource sector, the local small business sector was especially hard-hit. With job losses in the resource sector came fewer consumers, transportation improvements allowed out-shopping, and the rise of internet and online shopping also meant fewer local purchases. In addition, large resource companies focused on their increasingly tight profit margins could no longer simply go with a buy-local approach to needed goods and services, and instead started to contract out to low-cost bidders, regardless of whether they were local or not.

Having developed and thrived in the shelter of the local large industry, small businesses were now caught in the same cost-price squeeze as the resource sector. Despite a number of successful joint ventures between local businesses and larger businesses in distant centres, small businesses are still challenged to engage in this new era of industrial development due to the extent of operating capital, labour, and scale of work required for resource-based industries. At the same time, industry's use of mobile labour and distant staging areas has undercut the potential for small businesses to realize economic benefits from resource development through the provision of housing and related goods and services.

Over time, two trends emerged. The first was downsizing of the local small business sector as some economic activities – banking and automobile dealerships are especially notable in this regard – pulled out of such places. This left many empty storefronts and empty retail blocks – something that fed a negative feedback loop around the impression of a lack of local shops that exacerbated out-shopping by local residents.

A second trend, and one developed more recently, has been a take-up in the local economy by innovative small businesses. In some cases, these businesses have brought in new models to provide goods or services in ways that better match the size and needs of the local market place. In other cases, they have been exploring new and emergent market niches in connection with wider efforts by the local community to support local economic diversification and enhance local economic resilience.

In concert with this second trend, there has been a more general adoption of flexible business arrangements. In some cases, there may be a new storefront business, while in others, they may be home-based businesses or those developed within a local small business incubator facility. In the case of home-based businesses, these may or may not be just serving a local market as e-commerce and e-transactions become more widely adopted. In this sense, the quality of internet connectivity of rural and small town places is becoming a critical pinch point for small business success. Even as some rural businesses look to improve their online presence (Sorensen, 2017), senior government strategies to reduce and close many postal services has undermined the business infrastructure and capacity in these resource-dependent regions (Connelly and Nel, 2017c).

Employees

It is difficult to construct a portrait of small business in resource-dependent rural and small town places as they are quite diverse in their form and focus. As a result, it is also difficult to construct a portrait of the small business employment sector. To start, however, we can pull together some of the summary observations about retail and service sector opportunities that are part-time, generally low-wage, generally with few benefits, and generally without union organization.

However, there are some small business sectors that are quite different. For example, a local branch of a chain food store, or a bank or credit union. In these cases, the organization of work in the small town is similar to any place where those larger firms operate. Typically, so too are the wages and benefits packages for employees and managers. As resource-dependent rural and small town places manoeuvre boom-and-bust cycles, though, these types of small businesses may find it difficult to explain human resource management issues to distant head offices that do not understand the pressures associated with rapid growth or decline.

For small businesses, generally, there remain difficulties to recruit and develop employees who are often attracted to higher industry wages. This tends to leave small businesses with younger staff who have limited experience and management skills. These small businesses also often lack sufficient financial resources to further invest in staff development and even succession planning strategies. Small businesses may also fear that any resources invested into worker training and development will be lost once the most qualified and experienced workers move on to higher-paying industry jobs (Polèse and Shearmur, 2006). This continues to impact their operations through their capacity to deliver programs and services.

Conclusion

This chapter has provided a description of the basic and contemporary strategic responses of industry and small business stakeholders in the post-war political economy of resource-dependent regions. In this post-war era, big business has generally transformed from private or state-owned companies to publicly traded corporations to help with access to large forms of capital. This not only transformed big business but also had significant and broader implications for resource regions as corporations were now driven by their responsibility to investors and shareholders as they encountered accelerated boom-and-bust cycles for resource commodities. Large resource-based companies were responding to pressures to remain competitive by strategically pursuing cost minimization, reducing risks through joint ventures, off-loading activities to sub-contractors, and globalizing their investments. The result of these trends, often concomitantly impacted by Neoliberal policy approaches, has been a loss of employment and weakened union structures. Despite these changes, the high-paying employment positions that remain in resource-dependent regions continue to generate a two-speed economy.

The economic world has also changed for small businesses. They have adjusted to pressures external to the communities where they operate, as well as to local pressures. As such, small businesses continue to act as a barometer reflecting the health and resilience of the small town economies of resource-dependent regions. Despite the additional roles of small business owners as community leaders and as stakeholders directly impacted by the resource development, changes in industry procurement strategies and use of mobile labour has changed the small business climate of these small places.

8 Senior government

Introduction

In the global economy, as international trade agreements have been signed and super-national organizations such as the European Union and the World Trade Organization have formed, there have been questions about the continuing role of senior governments in resource development. In our examination of resource-dependent rural and small town places, it remains clear that while senior levels of government are a crucial piece of the political economy of development and transition, the pressures affecting these senior governments must be included in an understanding of processes of change and continuity. In this chapter, we explore the political economy of senior governments as it pertains to resource-dependent places and regions in developed economies.

Before we begin, we need to recognize that not all developed economies will have the same constitutional construction of senior government structures and responsibilities. Depending upon the political system, there may be only a single national government and then local governments; or there may be a national government, an intermediate tier of provincial or 'state' governments (using the US model and nomenclature), and then local governments. This chapter simply uses the notion of senior levels of government in order to bridge these different contexts.

The chapter opens with an introduction to the historical role of senior governments in resource development. Generalized trends between the end of the Second World War and today are introduced. The chapter also reviews a long standing characterization of rural development and resource industry restructuring – that of the 'addicted' economy. The chapter then explores the implications of the transition from a Keynesian public policy approach to a Neoliberal public policy approach. Building upon Chapter 6, the contradictions embodied within current Neoliberal public policy challenges senior governments and resource-dependent regions alike. Deepening the contradictions within a Neoliberal public policy context are the changing expectations of the population – expectations driven by the demographics of aging and the uneven geographies of capitalism. The next part of the chapter looks at the challenge of resource-development revenues and state spending. In a section entitled 'Six Year Olds at the Candy Store', we discuss the limited and chequered history of resource royalty trusts or legacy funds. The final section of the chapter explores the implications of international trade agreements on the authority and public policy freedom of senior governments.

Generalized trends

This first section of the chapter presents an overview of some very generalized trends regarding the role of senior governments and the development of both resource industries and resource-dependent rural and small town places and regions. The period of interest is from the immediate aftermath of the Second World War to the present. As noted in earlier chapters, the war had mobilized industrial activity on scales never previously imagined. The application of Fordist-style assembly-line production, Taylorist labour segmentation, and global supply chains of raw material inputs had created industrial output that dwarfed anything in the pre-war economies.

The long boom

As the world turned to peace, and also to the Cold War, the lessons of war-time industrial production were widely applied. In heavily industrialized countries, production turned to both consumer goods and Cold War armaments. In the resource peripheries of many of these developed states, the lessons learned about the benefits of being a secure supply warehouse for the industrial core regions grounded senior government development strategies aimed at realizing as much economic and social benefit from this global opportunity as possible. The economic benefit was to be found in the comparative advantage economic model of supplying a raw material when the seller has some cost advantage over other suppliers into an industrial market. In states struggling to meet the post-war demands for jobs and services, the economic potential embodied in the massive industrial development of previously untapped or lightly tapped resource 'banks' was a very attractive alternative.

The social benefits were also attractive. With large numbers of soldiers being decommissioned, senior governments had to find ways to get them into jobs. This was pressing not only from a concern over civil unrest, but also because the pre-war experience of many of these soldiers was with the latter years of the depression. Resource industries provided an opportunity for large numbers of jobs to soak up the post-war surplus labour pool. Since most resource industry jobs at the time generally required few skills and limited training, it was an ideal opportunity.

But to mobilize an industrial-scale resource-development policy in peripheral regions required very focused senior government programming and policies. To start, resource peripheries in developed economies tended to have limited industrial and transportation infrastructure, and they tended to have relatively small populations. Opportunities for resource development, therefore, required massive investments (state and private sector) in order to realize those opportunities. As described in Chapter 5 with examples from British Columbia, and elsewhere, the extent of senior government public policy attention, coordination, and investment needed to support new industrial resource development was sweeping. In Finland, Tykkyläinen et al. (2017, pp. 90–91) suggest that:

> National-scale government policies became a major tool for organizing the presence of the forest sector in the resource periphery. The emphasis of the Finnish industrial, development, and social policies were in mobilizing human resources to increase the exploitation of forests and cultivable land. These policies had a dual aim of connecting the rural areas to industrial production chains to increase the provision of raw materials, and industrializing the rural communities.

These investments and public policy changes were occurring within two wider contexts. The first was the continuing influence of a Keynesian public policy approach that gave license to senior governments to plan economic development at a broad scale. Economic planning, state investment, and public policy changes to shape economic opportunity, were commonly accepted as part of the day-to-day tasks of senior governments. The second wider context is what would later become known as the 'long boom' – the 1950 to 1970 period of economic prosperity that prevailed in most developed economies. Hand-in-hand with expanded production and the higher employment that funded expanded consumption, the long boom helped to cover over some of the vulnerabilities and dependencies of resource-dependent regions and economies. As described in earlier chapters, the risks associated with being price takers in the export of low-value resource commodities were hidden by generally ongoing increases in demands and prices over the long boom.

The first encounter with restructuring

The second of the generalized periods we wish to highlight concerns from the late 1970s through to the end of the 1980s. It is this period which pivoted the fortunes of resource-dependent regions and economies from growth to challenge, and which set in place the various processes of accelerated globalization, public policy shifts, and the restructuring of industrial production in many sectors. The period also signalled a definitive end to the long boom and the start of a series of increasingly dramatic economic up- and down-turns that continue to the present. The 1970s to 1980s period also signals the initiation of economic restructuring and the replacement of Fordist-style production methods with more Flexible production regimes. As recounted in earlier chapters, job losses rather than job opportunities became the story in established resource-dependent rural and small town regions (accepting, of course, that some regions would have job growth as new activities came to them – such as the oil sands region of Alberta or the Pilbara region of Australia). Commensurate with the restructuring of industrial production came calls for the restructuring of public policy and the withdrawal of the role of the state in both the economy and the lives of its citizens. The Keynesian to Neoliberal public policy shift was set against a backdrop of the economic challenges of this period and the trend of declining relative state revenues from the resource sector.

As the transition unfolded, however, it was supported by the continuing, but now 'hidden', hand of Keynesian public policy. The preceding several decades of state investment had built a robust industrial and transportation infrastructure, which successive economic booms could continue to exploit. The development of health and education systems had created a human resource infrastructure that increasingly adaptive industries would require in order to be both lean and competitive, and thus successful. While politicians and advocates began the mantra of tax cuts and reduced investment, they did so without acknowledging that they were simply exploiting and running down the assets set in place by previous generations without investing their fair share for future generations.

The period, together with its contributing resource sector benefits (of jobs and taxes) and its leaner, less engaged, public policy approach, also set the foundations for the near-abandonment of resource-dependent rural and small town places and regions to the fluctuations of the global economy. Under Neoliberal public policy approaches, senior government policy discourse argued that rural and small town places were now to be 'enabled' to play a larger role in defining their own economic future and finding ways to supplement or replace the economic contributors of historically important resource industries. As with the economy more generally, the discourse on enabling was cover for the withdrawal of senior government investments and interventions at the local level.

In New Zealand, Connelly and Nel (2017c) describe in detail efforts made in resource-dependent regions to become more entrepreneurial and to develop local or regional initiatives for marketing economic opportunities. These have included both community-based and regionally organized development associations and corporations. The approaches, and relative impacts, of such initiatives over time have been uneven. The lack of support from the state, has been telling:

> The state has since removed support such as the Community Employment Group (CEG) and other supports such as the Community Organisations Grants (COGS) have been reduced with some funds being transferred to the Department of Internal Affairs (DIA) to support three-year duration community development positions or activities in a few localities nominated as requiring assistance (Department of Internal

Affairs 2015). A very small amount of assistance was, until very recently, given via the Ministry of Social Development (MSD) to a few selected isolated and deprived rural areas to partially fund a community support worker to assist communities access to government social welfare services, however these have recently been closed due to a fatal shooting incident in one centre, resulting in the cessation of the work of eight support workers in Southland (Ministry of Social Development 2014). Minor funding is also given to REAP (Rural Education and Activity Programmes) in rural centres to help rural communities access parenting and community education support. ... Hence current state support for economically struggling rural communities in New Zealand is virtually non-existent, with the state focusing instead on growing regions and high growth potential business sectors, reflecting New Zealand's now long term acceptance of its neoliberal policy.

(Connelly and Nel, 2017a, p. 321)

This period has left some long-term challenges for resource-dependent places and regions. These include reduced investment in critical economic and transportation infrastructure and reductions in government spending on the local services needed to attract and retain both economic activity and residents. It also marked the start of a physical disengagement with resource-dependent regions, as senior governments sought to address budget challenges by closing down their regional offices. This process left centralized bureaucracies with a lack of 'eyes and ears on the ground' in rural and small town places – as a result, senior governments lacked the early warning signals of change (Markey et al., 2012).

But these stories do not capture the breath of experiences in resource-dependent regions. In the European Union, for example, development of the LEADER program has been a step towards supporting regional development in 'peripheral' regions. As noted in Finland, these supports can be a significant tool in advancing place-based development and revitalizing the services and infrastructure needed to retain and attract people, investment, and businesses. In reflecting upon the eras or phases of rural development, Kotilainen et al. (2017, p. 303) write:

Starting from the late 1980s, the third phase consisted of development actions based on specific development programs. While in the early phases it was the national government, ministries, and state-run development organizations that were responsible for advancing regional development, in the latter phases this responsibility was transformed to the regions themselves, with the regional councils becoming the chief organizations planning for regional development. The increase in the weight of regional organizations in regional development was also due to the membership of Finland in the European Union (EU) in 1995, which brought with it the pivotal role of regions in the allocation of EU structural funds. With Finland's membership of the EU, additional instruments were introduced into the regional policy repertoire. These included the EU structural funds and LEADER (from the French *Liason Entre Actions de Developpement l'Economie Rurale*) policy measures, which is a local development method seeking to mobilise local rural actors on the European scale (European Commission 2015). Therefore, over the decades, there has been a shift from supporting large scale manufacturing industries to an emphasis on competitiveness of regions in the global economy.

The age of uncertainty and unevenness

The third generalized period we wish to highlight is from the early 1990s to the present. The long boom and the first encounter with restructuring have been replaced with ongoing uncertainty and unevenness. To start, the global economic context for senior governments has been marked by successive booms and busts – each with seemingly different focal points and driving forces. Against that backdrop, calls for the expansion of Neoliberal public policy approaches to allow the economy to 'take off' during economic booms and to retrench senior government spending in periods of economic downturn had become even louder and insistent. Even as the 2008 global economic recession was noted as an exemplar of the excesses of an unregulated market, and where senior governments had to apply dramatic Keynesian public policy interventions to prevent the economic collapse from becoming a full-scale global depression, the Neoliberal public policy agenda was barely weakened.

Just as with the immediate post-war period, the age of uncertainty and unevenness is looking to the expansion and growth aspects of capitalism. For resource-dependent regions, senior governments are looking to the expanded industrial capacity of China as a place to sell more or new resource commodities, and to the energy transformation in places such as Japan, South Korea, and China in which to sell even more natural resource commodities. The economic model of the long boom is being re-enacted, but this time under a Neoliberal public policy approach such that opportunities are rarely realized and lagging resource-dependent regions continue to languish.

Addicted economies

As noted above, the success of the industrial resource-development model in the long boom period after the Second World War was quite remarkable. As industrial production expanded, so too did the demand for raw material inputs. As the wealth of developed economies grew, the burgeoning middle class and their baby-boom children became increasingly lucrative markets for that industrial productivity, and the cycle spiralled upwards for close to 30 years. While investments were needed in order to support industrial resource development in peripheral rural and small town regions, those investments would pay off over decades as resource industries and their related large cohorts of resource workers helped power not just local and regional economies, but sometimes even state economies as well. In countries such as Finland, Sweden, Australia, New Zealand, and Canada, all with under-developed industrial manufacturing bases beyond the resource sector, resource industries came to form the backbone of the post-war economy. As noted by Halseth (2016, p. 110), "Entire jurisdictions can become dependent on the wealth that natural resource production can generate – both from direct revenues and indirectly through taxation of workers and related spending".

Freudenburg (1992) long ago labelled the process and consequences of a public policy approach focused on resource exploitation, and on economies increasingly dependent upon the export of those resource commodities, as setting the stage for 'addictive economies'. The problems created by 'addicted' economies have been part of a wider debate – inclusive of our earlier discussion on Staples theory – about the potential of a 'resource curse'. A recent study by the International Monetary Fund highlighted a clear negative relationship between resource revenues and total domestic (non-resource) revenues. It also highlighted negative relationships between resource revenues and general diversity within the tax base. In other words, states with significant resource revenues had underdeveloped other aspects of their economies and had shifted the tax burden towards resource industry dependence. A further aspect of the resource curse debate relates to the artificial inflation that resource wealth creates with respect to currencies, exchange rates, and costs / prices / wages. Using the case of Canada, Polèse and Simard (2012, p. 23) describe how these national challenges are replicated at local and regional scales as well:

> As in national economies with rich resource endowments where resource rents drive up exchange rates, so in local economies resource specialisation may drive up local wages. In local economies, high wages replace high exchange rates as the first conduit by which competitivity is impacted. We found a strong positive relationship, controlling for other variables, between wages and specialization in resource extraction (drilling, mining, forestry, etc.) and in primary resource transformation (paper mills, aluminum and metal ores smelting, etc.).
>
> … [These pressures] in turn reduce competitiveness (specifically, in other export sectors), crowding-out such industries, depressing overall labour demand, and jeopardizing long-term growth. All three effects are visible in the resource-dependent Canadian urban areas, but with varying intensities.

As a counter to some of these tendencies, a recent report pulled together what the authors describe as the 'nine habits of effective resource economies'. These include:

- Saving resource revenues into robust resource trust funds,
- Adding value to the extractive sector,
- Continuing to invest in research,
- Maintaining global leading environmental policy,

- Maintaining a watch on emergent global trade opportunities,
- Investing in endogenous firms,
- Paying attention to trends in a regional and resource governance,
- Paying attention to developing the labour force for the long-term needs of the economy, and
- Developing natural resources within a national strategic framework.

(Drohan, 2012)

As noted above, during the long boom, the challenges and pitfalls associated with such an addiction to resource exploitation and export were largely covered over by ongoing growth in both consumption and production. However, as the global economy turned to its first encounters with economic restructuring, those challenges tested national economies and the economic viability of resource-dependent regions. In such cases, for senior governments there were immediate needs. The very large revenue flows from the massive volumes of natural resources that were being exported had to be replaced. But new economic sectors took time to develop and few seemed to hold the promise of natural resource wealth. As a result, there were concerted efforts by senior governments to find new ways to keep their resource sector economically viable as the global economy moved into the age of unevenness and uncertainty. These efforts would include reducing regulatory burdens, reducing costs via industry tax reductions or exemptions, and assisting with labour flexibility by changing employment legislation. All of these were focused on keeping resource revenues flowing into senior government accounts.

As we moved deeper into the age of unevenness and uncertainty, these older patterns and responses persisted. When resource-dependent places, regions, and national economies experience dramatic reductions in resource commodity exports, there are renewed scrambles to replace revenues. However, the lure of potential future large revenues (and wages, and taxes) from resource commodities within these addicted economies can act as barriers to effectively breaking from the past with new development directions or mixes of economic development at local, regional, and even national levels. As new resource commodity opportunities arrive, these addicted economies grasp onto the hope for a new 'gold mine' of resource revenue without considering how much the global economy, and their own senior government public policy options, have changed. As a staples producer in the global economy, Bowles (2013, pp. 261–262) describes how the BC state continues to re-run the 1950s development approach and actively works to "reposition itself within the multi-centred global economy through the large infrastructure projects now being undertaken across the region" to support new resource development – 1950s responses in the 21st century.

Neoliberal public policy and its contradictions

As described in Chapter 6, the public policy transition from Keynesianism to Neoliberalism has never been fully completed. In both theoretical and practical terms, there are numbers of inherent internal contradictions with a Neoliberal public policy approach, and these are very apparent when we look at changes and responses in resource-dependent rural and small town places and regions over the past 20 to 30 years.

In terms of resource development, many senior governments have used a Neoliberal public policy approach to 'open' the arena to multi-national capital to invest in their jurisdictions. But at the same time, senior governments are also employing a Keynesian public policy approach by manipulating market competitiveness through adjusting tax and royalty payments, by subsidizing the training and education costs for new industries, and by reducing the procedural barriers faced in various approval and regulatory processes.

Through the early 1990s, the BC government introduced two key changes to transform forest industry activity in the wake of sharp environmental criticism. The first was to introduce a 'high-stumpage regime' – which means increasing the royalty that forest companies paid to the state for each tree harvested. According to economic logic, as the resource became more expensive, existing forest product firms should have responded by moving from low value-added to increasingly higher value-added products – the companies did not, they just complained that the extra costs made them less competitive globally. The second was to introduce a new Forest Practices Code that would enhance attention and monitoring around forest harvest practices with an eye to enhancing environmental sustainability. Companies would take on more responsibility for the new requirements around planning and monitoring tasks. Industry also complained about these additional costs. Hayter (2000, pp. 351–352) describes the aftermath:

> In response, the Minister of Forests in June 1998 announced that stumpage levels would be reduced (by the year 2000) and the Forest Practices Code modified. Indeed, the Minister of Forests claims that, in addition to lower stumpage (amounting to $5 per cubic metre on the coast), the paperwork required would be cut in half … The total impact of these changes on industry is claimed to be savings of $300 million in stumpage by 2000 and $300 million in red-tape reduction by 1999. For industry, the changes to the Code are seen as a step in the right direction. They mean reducing the number of plans that have to be submitted (from six to three); eliminating some review processes; minimizing circumstances in which approved logging plans can be reversed; and allowing foresters to exercise more judgement in the field. For government, these changes are modifications but not reversals of policy. For environmentalists, however, this relaxation of regulations will gut the intent of the Code. Their particular concern is that undermining procedural requirements would modify the proactive approach to environmental problems implied by the Code to a reactive approach in the 'new' Forest Practices Code.

But not all states have vacillated in their policy responses with respect to the relationships between economic development and environmental protection. In Finland, for example, the forest industry in its various forms continues to play an important role in the national, regional, and local economies. There is also widespread recognition that the marketability of Finnish forest products depends a great deal upon how the global marketplace views their steward-ship of this natural resource and the environmental habitats that it maintains. At an industry level, approximately 90% of Finnish productive forests are covered under industry-led forest certification programming. Further:

> As a reaction to this changing demand, legislation and forestry practices were also modified to support more sustainable forestry and to make the origin of the wood transparent to purchasers. The Finnish Forest Act (1996/2014) requires safeguarding the biodiversity of forests by protecting the characteristics of particular valuable habitats in forestry operations, leaving uncleared timber zones, and considering environmental protection objectives. According to Similä *et al.* (2014) there have been relatively few identified breaches regarding the habitat regulation aspects of the Forest Act. Uneven-aged forest management, in contrast to silviculture practices of nearly coeval cohort groups of trees, has been possible since the 2014 amendment in the forest legislation, which increases biodiversity and improves the use of forests for recreation.
>
> (Tykkyläinen et al., 2017, p. 97)

While assisting at one level, the Neoliberal public policy approach for some senior governments has countermanded that assistance when it comes to not equipping rural and small town places in resource hinterlands to be ready for new industrial development when it comes. Long-term reductions in services and lagging investments in infrastructure are two challenges that industrial capital faces as it tries to bring investment and workers to the resource hinterlands of developed economies.

Another area of challenge for senior governments, and another arena for contradictions in the Neoliberal public policy approach, has been with respect to the expectations of citizens – especially urban citizens supported by urban media – around government services. This acts to create a tension between the state's desire to reduce government expenditure and step away from regulating the market and private business, and increasing calls from the public for the state to step in and manage market excesses or address concerns about business sectors.

While many urban areas may have geographic access to senior governments and to media, rural and small town places in resource-dependent hinterland regions are often very much out of sight and out of mind. In urban areas, citizens may be concerned about 'green' rapid transit options, while in rural and small town places, citizens may need to ride a bus for several hours each way along a mountain highway to access medical check-ups, due to the reductions of rural services and the lagging investment in telemedicine and telehealth options.

'Six-year-old at the candy store'

To this point, the political economy of senior governments with respect to resource development has focused on the opportunities for resource exploitation in the long boom period that followed the Second World War, the success of which helped embed resource development into national economies (and perhaps even addict those national economies to resource commodity revenues). The consequences of dependency have been clear even after the impact of the post-1980 recession with the way continuing economic up- and downswings impact resource-dependent regions and economies. So with continued focus upon natural resource industries following a 1950s industrial resource model, the question becomes: To what purposes have senior governments put their historical resource wealth in order to assist with periods of upset or longer-term processes of transition? The simple answer is that few senior governments have done well in terms of husbanding their historical natural resource wealth to help out in these contemporary challenging times.

Most households have heard the economic advice to 'save for a rainy day'. Over several generations, the collective wisdom was to spend a little and save a little, so that if sudden economic catastrophe befell the family, there would be the financial resources to help out. Unfortunately, many senior government jurisdictions have not heeded this message. Instead, they spent decades during the post-war economic boom spending the resource wealth that flowed into their general government coffers as fast as it came in. In many cases as well, they would act like a six-year-old in a candy shop – 25 cents in hand, the child would not only try to spend it all, but would try to talk the parent into giving them just five more pennies so that they can buy another treat. For senior governments, this is known as running a deficit. Over time, the 1970s experience with running small deficits to mediate short-term challenges with revenue flows became a habit that grew, that was hard to break, and ended up putting considerable debt loads on many senior government budgets. The only saving grace for contemporary governments is that historically low interest rates over the past two decades have kept debt servicing costs viable for many of these jurisdictions. If interest rates were to rise dramatically, however, the debt exposure of many states would create significant global problems.

In a jurisdiction like British Columbia, the six-year-old in the candy store scenario has been played out over and over. The forest industry, for example, has been a significant part of the economy for over 100 years. Yet, not one penny of the revenue of those 100 years has been put into a trust or legacy fund. Instead, all of the revenue goes into senior government general revenue accounts for spending in the fiscal year it was generated. From this, the state did make significant long-term investments in infrastructure across the province such as highways, schools, hospitals, etc. But, after 100 years of sustainable resource development being harvested and exported, there is no monetary legacy from the forest industry – no monetary legacy if the industry ended tomorrow, and no monetary legacy to continue to fund the renovations, renewal, and replacement of the infrastructure developed years and decades earlier.

BC has been even worse when it comes to non-renewable resources. With the advent of fracking technology, new natural gas reserves were charted in British Columbia. As discussed earlier, BC followed the path of selling lease rights to these reserves to multi-national companies. The lease rights were a one-time revenue source. If developed, the province would also see royalty and tax benefits, but those benefits would pale in comparison in the windfall reaped in the sale of leases. However, the billion-dollar revenues that came in over a short number of years from those natural gas lease sales all went into general revenue and helped senior governments balance their budgets as part of an ongoing political dialogue in BC about which political party is a better 'manager' of the provincial budget.

For states that did not save for a rainy day, and which have not only seen the strain of Neoliberal public policy approaches on infrastructure and service expenses in resource-dependent regions, but which had experienced a growing backlash from these same regions that had otherwise supported massive revenue flows to the central state, there have been experiments with new resource revenue transfer programming. One of the most noted of these resource revenue transfers experiments is Western Australia's 'Royalties for Regions' program.

Royalties for Regions

In 2009, Western Australia initiated its Royalties for Regions initiative to promote community and economic development in resource producing regions (Haslam McKenzie, 2013). Through this program, 25% of annual royalty payments received from industry were to be reinvested through three funds, including the 'Community Local Government Fund', the 'Regional Community Services Fund', and the 'Regional Infrastructure Fund' (Argent, 2013). Revenues from royalties have also been invested to support the 'Pilbara Cities' program, that is designed to increase economic and industrial diversity, assist with planning for growth, invest in a breadth of projects to improve services, and coordinate infrastructure development (Franks, 2012). In fact, a state planning and development framework was established in 2013 to coordinate investments from the Resource to Royalties program with other department programs (Paül and Haslam McKenzie, 2015). Investments were also made in cross-regional projects to strengthen broader infrastructure and coordination (Tonts et al., 2013). While this is limited to Western Australia, other Australian states have experimented with such initiatives. For example, in Queensland, mining companies also pay special mining rates to support services in rural outback locations (Anglo American Services, 2012).

The Royalties for Regions program, however, has been criticized for the limited use of a strategic vision to support investments towards the long-term growth, diversity, and resilience of resource-dependent places and regions (Rolfe and Kinnear, 2013). Limited strategic analysis of mobile workforce issues can misdirect royalty revenues from where such funds are most needed (House of Representatives, 2013). Royalty revenues are also increasingly being shared among communities outside of resource-dependent regions, despite acknowledged gaps in the physical and social infrastructure in rapidly growing resource-dependent communities that are required to support industry and broader economic growth (Kelsey et al., 2012). In one study, for example, stakeholders suggest that during the first round of funding, the Royalties for Regions program was open to 14 local governments impacted by rapid resource development (Ryser et al., 2015). During the second round, the competition for funding was opened up to 30 regional councils that were either directly or indirectly impacted by industry growth. Some have even suggested that regional development councils have become dependent on the Royalties for Regions program (Paül and Haslam McKenzie, 2015).

The Royalties for Regions program has also been criticized for reinforcing a staples economy through extraction rather than economic diversification. As Tonts et al. (2013, p. 372) state,

> A considerable amount of funding is being spent on upgrading transport infrastructure, including road, rail and the expansion of the port in nearby Esperance. All of this is focused on improving the efficiency of transporting raw materials out of the region, and thereby reinforces the truncated economic development that Innis described … More telling is the Exploration Incentive Scheme that is funded as part of the Royalties for Regions. This provides around A$80 million per annum to assist companies to explore for new mineral resources.

Not only are Royalties for Regions funds potentially subsidizing industry by returned monies to support actions that industry would otherwise have had to spend as part of project development, but the same may be the case when Royalties for Regions funds subsidize or fund senior government activities that they also might otherwise have had to undertake as a normal part of their activities. Senior government programs, for example, have also been using royalties and other industry-related revenues to strategically target investments in regional growth centres (Government of Western Australia, 2010).

Building upon our theme of addicted economies, senior governments can become dependent upon resource royalty schemes and funds to support their investment and revenue strategies without considering the complications of intervening resource commodity boom-and-bust cycles. In British Columbia, the provincial government announced its plans in 2013 to develop a new Prosperity Fund that would be directly tied to the establishment of BC's LNG sector; a sector that was anticipated to generate $1 trillion in cumulative GDP over 30 years, with more than $100 billion in anticipated revenues for the Prosperity Fund (Office of the Premier, 2013). Contributions to the fund were supposed to come from royalties, sales tax, and corporate income taxes collected from industry (Hunter, 2016). Many LNG projects, however, have been postponed and face uncertainty due to poor market conditions. The provincial government has been criticized for putting all of its eggs in one basket, with many worried that the anticipated benefits will never materialize. As the media have recently reported, "The BC government is set to make the first contribution to its long-awaited prosperity fund in Tuesday's budget, but the $100 million going into the fund is not linked to LNG revenues" (CBC News, 2016). Instead, the contribution will come from an anticipated surplus from general taxpayer revenues.

There are many other jurisdictions that encountered a period of resource wealth and did not create a legacy fund, i.e. did not put a penny away for a rainy day, including the Netherlands and the UK with their North Sea oil and gas fields, Oregon's forest economy, and others. But not all senior governments proceeded in this short-term manner. Some created legacy or trust funds for the future. In the paragraphs below, we touch upon several examples.

Royalty trusts

Resource trusts or funds are critically important for senior governments. They provide a way to save and grow excess revenues from economic upswings to help cope with periods of economic downswings. If accumulated incrementally, they can be a stabilizing force in the state economy and can provide the economic support needed for structural adjustments in that same economy. As noted, not all state governments have adopted a trust or fund savings model. In jurisdictions dependent upon non-renewable resources, this is especially problematic. In this part of the chapter, we review several examples of resource trusts, the size and character of which vary tremendously.

The long post-war economic boom in developed economies supported many remote, rural, and small town regions through the establishment of resource development. Many jurisdictions have, and continue to process, resource-development proposals though special state-industry (and increasingly) – community / region 'agreements' (Horsley, 2013). Over the past 60 to 70 years, senior governments have had opportunities to save some of their resource incomes into trusts or other 'rainy day' funds. Motivations for the development of such resource trust funds can come from different sources. In BC's Peace River region, an area of extensive natural gas extraction, "municipal actors within the region were able to forge a coalition and craft a negotiating position and agreement that superseded traditional barriers associated with parochial local interests" (Markey and Heisler, 2011, p. 60). The 'Fair Share Agreement' reallocates a portion of provincial oil and gas revenues back to municipalities. Motivations for rural communities to work together come from a history of exploitation. In the Australian context, Tonts et al. (2013, p. 365) note that throughout "the nation's previous resources booms, there were marginal transformations in the economies of those regions from which the minerals were extracted. Overwhelmingly, the wealth generated by the resource boom has concentrated in the cities". If dependent upon the local motivations, however, the pace and scale of contemporary resource-development booms can significantly limit efforts to develop long-term benefits. As noted by Haslam McKenzie (2013, p. 341), "community engagement strategies and the collaborative planning processes have been undermined by disconnects between commercial imperatives, government frameworks, investment risk and timeframes". A current source of pressure comes from the need to renegotiate the relationship between Aboriginal / Indigenous peoples and the state and industry actors across traditional territories (Trigger et al., 2014).

While not all senior government jurisdictions have created such resource trust funds (Davis and Tilton, 2005), for those who have, the characteristics of the savings and spending plans associated with such trust accounts reflect something of a social, political, and economic character of the state. This section briefly introduces three such examples.

Alaska Permanent Fund

The discovery of oil reserves in Alaska's Prudhoe Bay and other parts of its North Slope region in 1968, and the subsequent construction of the Trans-Alaska Pipeline to the port of Valdez on the Pacific Ocean, created a new economic opportunity for the state. After much debate, the Alaska Permanent Fund (APF) was created in 1976 as a means of providing stability and sustainability by allocating oil resource revenues into a savings and investment fund, so as to support future opportunities for growth and prosperity (Community Development Institute, 2007). The idea for the Fund was germinated in 1969, the year Alaska auctioned drilling rights to 164 tracts of state-owned land at Prudhoe Bay (www.apfc.org/home/Content/home/index.cfm). The proceeds amounted to US$900 million. This sparked debate about what should be done with the money. Proponents of the Permanent Fund argued that:

- the fund could help to create an investment base from which to generate future income even after oil revenues have diminished;
- the fund could remove a significant portion of the revenues from government, reducing the opportunity for excessive spending; and
- the fund could take non-renewable wealth and transform it into renewable wealth.

In the end, the Fund was created through a statewide referendum and enabling Constitutional Amendment. The Alaska Permanent Fund is made up of two parts: the Principal and the Earnings Reserve. The Principal is the dedicated part of the Fund and once monies have been allocated to the

Principal, they cannot be removed unless approved by a majority of all voters in a statewide plebiscite. Sources of the Principal include:

- dedicated oil revenues automatically deposited in the Fund under the terms of the State Constitution; the legislated amount is 50% of all mineral royalties (25% prior to February 1980);
- legislative appropriation; and
- income transferred from the Earnings Reserve to provide inflation-proofing.

The Earnings Reserve is an accumulation of net income from investment earnings. The Fund invests in a number of financial assets and real estate in order to generate income and capital gains. All investments follow the 'Prudent Investor Rule', where the security of the principal outweighs the risks of high returns.

The Permanent Fund is operated as a trust and its portfolio of investments is managed by the Alaska Permanent Fund Corporation – an agency at arm's length from government. It is accountable to the public indirectly through elected representatives, and has legislated public reporting requirements. Investment strategies are outside the control of the government.

In practice, the emphasis of the Alaska Permanent Fund is on generating financial dividends for residents. With Alaskan residents receiving direct dividend payments, it is difficult to introduce changes because individuals see tangible benefits 'in their pockets' rather than less tangible long-term benefits that are more commonly derived through infrastructure and community / economic development investments. This may not translate into lasting benefits for communities.

Alberta Heritage Savings Trust

The Alberta Heritage Savings Trust Fund (AHSTF) was established by the Alberta provincial government in 1976. It was an attempt to diversify the economy, and it was hoped it would deploy social dividends, along with financial returns (Community Development Institute, 2007). Four background arguments emerged for supporting the Alberta Heritage Fund:

- supporting fairness to future generations through recognizing that Alberta's non-renewable resources were depleting;
- supporting substantial future capital investment so that the AHSTF can assist with mitigating some of the challenges associated with economic restructuring;
- providing quality-of-life improvements that Alberta could not otherwise afford; and
- providing an alternative revenue base for the future.

(Warrack and Keddie, 2002, p. 4)

The AHSTF was established with a purpose of saving for a time when natural resource revenues would begin to decline. It was hoped that during boom times, the economies would have the opportunity to diversify, while in the meantime a 'nest-egg' would provide security and stability for the future (Warrack and Keddie, 2002).

While the AHSTF was to function as both a savings vehicle and an economic policy lever; the latter objective came to dominate. In Alberta, it was argued that by diversifying the economy through public policy, the future of its citizens would be better protected against the volatility of a natural resource-based economy. Eventually, the economic diversification objective gave way to the Alberta government's desire to reduce its need to borrow from capital markets. As a result, the funds came to be used by the Alberta government for financing budget expenditures.

The Alberta government provided the AHSTF with an initial instalment of CAN$1.5 million in 1976. Through the first decade, a portion of revenues from natural resource royalties was also allocated to the AHSTF, and investment yields were ploughed back into the AHSTF until 1982. Like in Alaska, the 'Prudent Investor Rule' was applied. In its first approximately 30 years, the AHSTF benefited Albertans in that more than CAN$28 billion in investment income had been transferred to the province's General Revenue Fund to support program spending in areas such as health care, education, infrastructure, debt reduction, and social programs.

However, unlike Alaska, operation and planning of the AHSTF remained the responsibility of the Provincial Treasurer. As a result, the government has wide discretion as to where the funds will be directed. Although the Treasurer must provide annual reports on investments and performance, accountability remains primarily within government. Critics argue that the Heritage Fund is nothing more than a political lever used to implement and reinforce public policy decisions. They allude to the fact that income is transferred directly into General Revenues, indicating that the direction of the fund is dependent on the desires of the government of the day.

The deposits and the use of investment revenues have not been as disciplined as other trusts. Priaro (2015) reports that in 1987, the value of the AHSTF was estimated at CAN$12.7 billion. By 2014, its growth was only to approximately CAN$17.5 billion.

Norway

As described earlier in the chapter, as the Norwegian government began to mobilize its North Sea oil and gas opportunity, it chose to do so as an industry actor, not just via the sales of leasing rights and subsequent taxes. The Norwegian state established its own oil and gas company – Statoil ASA, which came to be headquartered in Stavanger, Norway. Established in 1972, Statoil was merged in 2007 with the oil and gas division of another Norwegian state energy company, Norske Hydro. Today, with operations in more than 30 countries worldwide, Statoil is a major global supplier of oil and gas.

As profits from the North Sea oil and gas projects of Statoil started to flow in, the Norwegian state government established the Petroleum Fund of Norway. By legislation, a set percentage of annual profits would flow into the fund, only the interest can be spent, and even then, its dollar value is limited to a set amount. The purpose behind the fund and these disciplined rules was to establish a legacy for their non-renewable North Sea resource endowment, and to avoid some of the challenges that resource wealth, and then the collapse of that wealth, had on other state economies, and finally to help smooth out some of the fluctuations in global demands and prices that is so common in resource commodities such as oil and gas.

Today, the Petroleum Fund of Norway has been renamed as the Government Pension Fund Global. As noted, it has grown to a significant size (in the order of US$900 billion). The effectiveness of the fund has also been shown in that the Norwegian state barely noticed that there was a global economic recession that started in 2008. As the fund moves forward, the Norwegian state is trying to be responsive to the wishes of the populace. For example, growing concerns about ethical investments (an interesting turn of events for monies earned from oil and gas developments) has prompted the fund to undertake a divestment strategy and to redeploy its assets.

International trade agreements

The last part of the chapter looks at the issue of globalization and its impacts on senior governments through the lens of international trade agreements. In particular, two forms of international trade agreements are considered – longer-term process agreements and more recent bi- and multi-lateral trade pacts. In both cases, the implications include some level of sacrifice of state autonomy as part of gaining additional access to a market or markets. It is the trade-off involved that generates the debate.

To start, international trade agreements are not an especially new concept, even if the terminology and details might differ. The question of international trade agreements, however, took on renewed urgency through the Second World War and its immediate aftermath. The depression had created the fertile soil for unrest and nationalist movements that seized governments in numbers of states. The collapse of empires was feared to bring further uncertainty with the result that international discussions to calm financial and currency markets, and to set the foundations for rebuilding war shattered economies, began. As noted in Chapter 4, one of the first of these was the 1944 Bretton Woods Agreement that established the US dollar as the world's *de facto* currency benchmark. From this base, other institutional actors would also emerge including the International Monetary Fund and the World Bank.

To support the longer-term management of the global economy, processes for managing international trading relationships were needed. One of the most important of these processes has been the long-running negotiations around the General Agreement on Tariffs and Trade (GATT). The first round on GATT negotiations, managed under UN guidance, was concluded in Geneva in 1947. The purpose of the GATT was the regulation and management of global trade through negotiations towards the elimination of trade barriers and related preferential trading alliances. Over the decades, there have been successive 'rounds' of GATT negotiations.

At the Uruguay Round of GATT negotiations, which ran from 1986 to 1994, a new platform for global trade negotiations was established – the World Trade Organization (WTO) – in 1995. As an institutional body like the IMF and World Bank, the WTO had extended its coverage to a wider array of trade sectors and topics than had the GATT.

Aside from the framework for globalization that the aforementioned processes worked to create, there has been increasing interest in the establishment of specific country-to-country international trade agreements. As noted in Chapter 4, these might include bi-lateral or multi-lateral agreements, such as the North American Free Trade Agreement (NAFTA) or the more recent Trans-Pacific Partnership (TPP). Such trade agreements have a spatial organization or a spatial logic underlying the partnerships, and they typically focus on matching access to manufacturing markets for the raw materials from less-developed states and access to populous markets for more-developed states. As noted, there is also a political element with states giving up some of their sovereignty in order to 'ease' the flow of people, money, materials, goods, or services within the trading block.

As a part of Neoliberal political economy landscapes, then, international trade agreements have opened up markets to low cost production regions and have enabled the increasing movement of capital and mobile labour in global resource regions (Vodden and Hall, 2016). They have also provided the institutional foundations to support more flexible modes of production and strategic joint ventures among global partners. These agreements, however, have also introduced the mechanisms and structures through which the investment capital is protected. As such, these arrangements are serving the imperatives of capital and shareholders while remaining quite silent on broader implications for resource-dependent rural and small town places. For senior governments the implications are complex and include both positive (opportunities to grow their internal economies) and negative (increasing vulnerability for some of their internal economies) possibilities.

Conclusion

The shift from Keynesian to Neoliberal public policy frameworks has transformed senior government approaches and responsibilities in resource-dependent regions. During the post-war period, senior governments made critical infrastructure and service investments that would position their hinterlands for community and economic development success for decades. Reinvestments in transportation and communication infrastructure, research and innovation, and the resource-based communities themselves became much more limited in the 1970s and 1980s as successive governments chose to run down previous assets. Broader Neoliberal pressures affected senior government strategies and limited both the responsiveness and effectiveness of their policy and governance approaches to better position their resource-dependent regions. Under Neoliberal public policy approaches that removed many senior government offices and supports, resource-based communities found themselves increasingly on their own to chart their own path forward. The reductions in spending and investments underscored any real capacity to support community renewal and transformation processes.

As the long post-war boom came to an end, and economies encountered frequent recessions within the cyclical boom-and-busts of their resource commodities, senior governments have struggled with the contradictions inherent in their Neoliberal public policy approaches that frequently required a return to Keynesian interventions. Regulations have been weakened in order to reduce 'red tape' and improve the fiscal positions and corporate interests in the global economy. As many senior government stakeholders continue to be addicted to resource-based economies, they have pursued new addictions afforded through royalty revenues, trusts, and the like.

9 Place-based communities and local government

Introduction

While the previous chapter covered the changing political economy of senior governments, this chapter changes the spatial scale – specifically, to place-based communities and their most typical form of organization: the local government. As sketched throughout the book, our focus is on resource-dependent places that are subject to the accelerating changes arising from globalization and from changes in public policy approach. This chapter is about defining and understanding those places within that context. The change in spatial scale is important, as it is 'on the ground' that the decisions of international capital and the changes in public policy are most keenly felt. Therefore, the 'local' provides a suitable scale at which to explore the impacts, and the contradictions, of the changing political economy of resource-dependent regions within developed economies.

This chapter is comprised of four sections. To start, the concept of communities, specifically place-based communities, is introduced. The focus will not be on an exhaustive review of 'what is community'. While such a question is important, and has occupied a great deal of social science literature over the decades, our focus is to local impacts of accelerating globalization. The second section outlines the role of local governments. This includes not just the historical context of local government structures, but the current trends and pressures forcing a rethinking of government and governance at the local level. The significant public-policy shifts from Keynesianism to Neoliberalism, and the fiscal crises of many senior governments, have created an array of challenges for local governments. Asked to do more around community and economic development, these local governments are constrained by their lack of access to critical policy levers or to investment decision making. The third section focuses upon community readiness and processes of change. The final section focuses on how small places can compete in the global economy. This includes consideration of the needs to scale-up local community interests to working with neighbours and partners in regional initiatives.

The local scale is one where the trajectories of historical development collide with the imperatives of the contemporary global economy. A theme running through the chapter is the degree to which local places and local governments have recognized the circumstances and opportunities of the new global economy or how they have languished trying to re-create the old rural economy of resource commodity exports.

Communities – a brief definition

As noted, the concept of community is both clear and complex. As a word, we use it every day, and we use it in an array of contexts. We use it to describe a neighbourhood or a locale where we might live or might have grown up, we use it to describe a sense of membership with a group of others, and we use it to describe a process of coming together when there is a challenge or a crisis that needs a collective response. While we use it every day, the concept of community becomes more complicated when we try to define it. That complexity comes from the fact that there are so many uses of the term that we end up with many different definitions and types of definitions.

At its simplest, community conveys some sense of membership or belonging. Donnermeyer et al. (1997, p. 67) define community as "a combination of people and groups which perform certain functions that are locality-based, including the production / distribution / consumption of goods and services, socialization, social control, social participation, and mutual support". While the academic literature that has attempted to define community is lengthy, and the debate around definitions has stretched back many decades, there are some areas of general agreement. These include that a community provides both the social and spatial framework for the day-to-day activities of individuals, that it is bound together by a shared sense of belonging, and that it identifies something of a distinctive identity for its members.

Belonging and identity are not unproblematic. In the 'oil sands' town of Fort McMurray, Alberta, the arrival of workers from around the world on short- through long-term work rotations creates a curious venue for exploring notions of 'community'. On the one hand, despite differences among labour migrants, "High wages in the oil sands well exceed the Canadian average, making complex class differences less apparent here than elsewhere. This lends itself to a homogenizing narrative of community despite differences in wages, background, citizenship status, and so on" (Major and Winters, 2013, p. 141). But not all are 'welcomed' under this homogenizing narrative. As described by Foster and Taylor (2013, p. 167):

> The rapid expansion of the oil sands in northern Alberta in the early 21st century led to the use of significant numbers of temporary foreign workers. These foreign workers became a part of the regions so-called 'shadow population' ... [They] are excluded from the life of the community due to their differential exclusion, vulnerable and precarious connection to the labour market, experiences of discrimination, and conflicted transnational community identities.

The process of constructing or reproducing a sense of community involves several important elements for our study of the political economy of resource-dependent regions. These include that communities play a role in processes of socialization and the transmission of local cultures. The norms, behaviours, and expectations of remote, rural, and small town communities are reproduced through these processes. Community also supports processes and opportunities for social participation and for the social interaction of members. This is where ideas and opinions are shared and formed. Depending upon the degree of openness within community structures, these two processes may act to reinforce older ideas and behaviours, or they may push members towards new ideas and new behaviours. In the fast-paced global economy, this question of mindsets – of being open to the future or of being closed and living in the past – is one of the most significant issues for resource-dependent regions undergoing dramatic change.

Top-down, bottom-up

As described in most introductory texts, there are two stereotypical ways by which definitions of community are constructed. The first involves top-down definitions. In this case, some form of jurisdictional or managerial boundary is imposed upon or drawn around a geographic territory. In the case of local governments, their territorial boundary line marks whether one is included within their jurisdiction or excluded. Many of the debates and discussions around the concept of community, and many of the public policy initiatives designed to assist with rural community development, are hampered by their deployment of purely top-down definitions. Among the key reasons for this is that for individuals, top-down definitions may be very far removed from the way people construct and carry out their day-to-day lives, and for functions of the global economy the constraints of a bounded local community fail to capture entirely within its borders any of the relevant processes.

In contrast, interest-based communities are described as being defined by the members themselves. Such definitions may emerge around interests or activities, but regardless of the focus there is something that acts to bring these individuals together in a way that meets some of the most basic defining characteristics of a sense of community. Attention to interest-based communities has always been an important part of resource-dependent regions as people become involved with different organizations or interests that connect within or beyond the jurisdictional framework of their top-down and place-bounded community setting. Attention to interest-based communities has grown commensurate with the rise of the internet and the construction of communities without propinquity. One of the challenges for interest-based community definitional frameworks is that researchers have compensated for the lack of a spatial reference base by describing very specific and discrete sets of criteria that can be 'checked off' in order to determine whether a particular network, group, or organization satisfies a particular definitional framework.

The use of such checklists distracts, however, from the important question of meaning behind such interactions and such constructions of community.

Between the two are a host of approaches and conceptualizations of community. Valentine (2001) provides a very good and detailed review of the history of these in her social geography textbook. Among the various topics discussed are 'natural communities' and 'neighborhood communities'. She also explores a range of interest-based communities, and considers ways by which the sense of community has been both romanticized and imagined.

From this point forth, and with our focus upon resource-dependent places, we will employ a definition of communities that is distinctly 'place-based'. It draws upon a combined definitional approach that includes the spatial reference points of top-down approaches and yet still maintains the transactional flexibility of interest-based approaches. For resource-dependent regions, we have found that this place-based approach is quite suited to capturing the range of purposes to which the concept of community is deployed.

With our interest in small resource-dependent communities, we are not just interested to define place but also aspects of small places. Each OECD country has its own definitional framework for 'urban' and 'rural' (usually with a host of gradients in between) as fits the context of the state's geography and demography. Internationally, governments and agencies define both rural and small towns differently in ways that reflect not only the circumstances and geographies of regions or countries, but also the mandates and missions of the agencies formulating those definitions.

Resource-dependent rural and small town places can be distinguished from other places by their physical isolation and dependence on global resource commodity markets that create dependencies and vulnerabilities. These places have been strongly shaped by the global capitalist economy and their role in staples production. The identities and meanings associated with resource-dependent rural and small town places, however, can be both contested and manipulated by a range of conflicts and interests (Davenport and Anderson, 2005). This may include, for example, the significance of natural resources to Aboriginal culture, or the significance of natural resources to the local economy and broader regional growth. In this way, the attributes of small places can also be shaped by their physical, economic, cultural, and social foundations. These attributes, or assets, are never static but are constructed and reconstructed through social, cultural, economic, and political processes (Cheng et al., 2003).

Communities and global capitalism

Because we are interested in the process of community, and its implications for community development, within resource-dependent regions, these communities also link to our conversations about labour and capital. To start, communities provide the setting for the reproduction of labour. As noted by DeFilippis and Saegert (2012a, p. 3),

Communities, whatever their form, are the realm in which social reproduction occurs. That is, communities are the sites for our housing, education, health care, daily convenience shopping, and other activities that sustain us physically, emotionally, socially, and psychologically. Labor is a peculiar commodity in capitalism in that, unlike other commodities, it needs to replenish itself when it is not being used, and is needed to produce more labor to be purchased in the economy in the future. A worker simply cannot continue to work without having time for sleeping, eating, and maintaining his or her health.

In the addition to being sites for the reproduction of labour, resource-dependent places and regions have also been sites for the reproduction of capital. As noted earlier, post-Second World War industrial resource development require the participation of multi-national firms with the size and scale of operations to support the significant investment needed to build new centres for the harvest / extraction and processing of resource commodities. Place-based communities in resource-dependent hinterland regions became the sites for those investments.

From the immediate post-war period until the late 1970s and early 1980s, both capital and these place-based communities benefited from the relationship. Capital had access to the raw resources, while communities provided for the home and services needs of the workforce. Capital and labour both contributed taxes and spending to help maintain community amenities, services, and infrastructure.

The significant changes in the global political economy after 1980 impacted the resource sector of developed economies quite significantly. As capital began to adjust to the new realities of an increasingly open and globalized economy, the communities that accommodated their investments and workforces were faced with a challenge. As DeFilippis and Saegert (2012a, p. 3) point out,

> Place-based communities are necessary loci for the functioning and reproduction of global capitalism, just as they have been for earlier historical forms of political economy. Yet, as the early sociologist understood, these communities increasingly do not themselves control or contain the forces of either production or reproduction. Communities are therefore in the contradictory positions of being vital for the maintenance of the larger political economy, but significantly constrained in what they can achieve in terms of shaping or transforming that economy.

This is not a new circumstance, however, as remote, rural, and small town communities have never been in a position to exercise power in the shaping of the economy or industrial investments. But this positionality relative to the pace of restructuring has created a disjuncture in responses – capital has generally reacted more quickly while communities have lagged.

In addition to being sites for both capital and labour, resource-dependent places are also sites for the ongoing process of globalization and restructuring. In the rest of this chapter, we pay attention to two aspects of these ongoing processes in particular – the unevenness with which resource-dependent regions are being impacted, and the suite of new development pressures and opportunities that are presenting themselves.

In an effort to 're-situate' resource-dependent regions in the global economy, we draw on the idealized characterization of a 'globalized countryside' as described by Woods (2007). He described the global countryside through ten characteristics (list adapted from Woods, 2010). These include:

- Primary- and secondary-sector economic activities that are dependent upon the demands of global commodity chains;
- Increasing corporate concentration and integration at a transnational scale;
- Serving as a supplier, and employer, of migrant labour;
- Tourist flows evident in sites of global rural amenity;
- Sites for offshore property investment;
- Commodification of nature, including global values of environmental protection;
- Inscription of the marks of globalization, including everything from industrial investments through to the presence of invasive plant and animal species;
- Increasing social polarization;
- New forms of governance and political authority; and
- Rural places serving as increasingly contested spaces.

When we employ such idealized characterizations, we need to recognize that we would not expect to 'see' all of the attributes of such characterizations in any particular place, but that it provides a framework for interpreting the particular impacts of globalization in resource-dependent remote, rural, and small town places.

By combining the global and the local, as well as space and place, such a characterization allows for analysis of individual rural places relative to the uneven pressures of globalization. That is,

> The extent to which any particular characteristic is evident in any particular rural locality is determined not only by the degree of penetration of globalization processes, but also by the way in which those globalization processes are mediated through and incorporated within the local processes of place-making.
>
> (Woods, 2007, p. 494)

Mitchell has examined the social and economic transitions underway in different small town places in Canada. In a 'city's countryside' example, she has traced the creative destruction process as productivist economies transform into heritage and tourism-based economies (Mitchell and De Waal, 2009). Following the logic of evolutionary economic geography, the transformations are identified as ongoing in response to changing stresses and opportunities, and that vestiges of past economies linger. In more remote resource-dependent towns, the complexity of transition is identified, as is the imperative of transition to more diversified and resilient economies (Mitchell and O'Neill, 2016).

Communities and change

In order to comprehend and then respond to the impulses of change (economic, policy, environmental, cultural, etc.) communities require robust levels of capacity. In describing community capacity, Sullivan et al. (2014, p. 222) argue that,

> The foundation of community capacity stems from traditions and social relationships between residents that facilitate routine interaction. Community capacity is also nurtured through responses to both positive and negative economic, social, political, and environmental stresses that often take the form of collective behavior (social cohesion), instituted through formal or informal networks of trust (social capital).

Breaking down elements of capacity into processes of social cohesion and social capital add additional complexity. As argued by Sullivan et al. (2014, p. 222),

> There are two main types of social capital: bonding and bridging. Bonding social capital occurs when trusting relationships are formed locally or within local groups, and intensifies local ties, strengthens the ability within local groups and organizations to work collaboratively, and addresses local needs and problems (Michelini, 2013). Bridging social capital occurs when trusting relationships are built inside or outside the community and links are made to a wider pool of ideas, experiences, advice, and support outside the regular circle of interaction.

They conclude that attention to building social cohesion, social capital, and the resulting community capacity at both local and regional levels, "can enhance the ability of stakeholders to improve the resiliency and stability of rural and small town places" (Sullivan et al., 2014, p. 225). Each of social cohesion and social capital function and interact recursively to build capacity and start the process of engaging that capacity in community development. For example, enhancing social capital access to networks can assist in providing a wider range of supports and resources. These additional supports or resources then help to enhance both social cohesion and social capital.

But the social cohesion and social capital aspects of community capacity are also supported by, and in turn support, a number of other community 'capitals'. Included in this 'multiple capitals' framework are human, cultural, social, political, financial, built, and natural capital (Emery and Flora, 2006). Human capital has always been an important element of community capacity. In general terms, it refers to the physical health, knowledge, skills, training, and experience (including experience through paid and voluntary work). All of these attributes can help improve individual well-being and capabilities. Cultural capital is also important, but has proved to be more challenging to define (Kingston, 2001). Again in general terms, cultural capital has referred to cultural traditions, language, social groupings, and the social norms that influence how people understand, structure, and experience their daily lives. As noted by Dumais (2002), cultural capital is especially important, as it can influence access to, and use of, other forms of capital. Shucksmith (2004) highlights that this is important in contexts where identities have been socially constructed and re-constructed through the processes of colonialism and assimilation. For Aboriginal peoples, especially, the processes of social, economic, and cultural marginalization have a continuing impact. As described above, social capital refers to the social processes and networks between individuals and groups that create and support opportunities to build trust, share resources and information, facilitate cooperation, and develop social skills. Political capital is driven by the influence or power that enables people and institutions to connect to, and control, a broader range of resources and political decision makers (Emery and Flora, 2006). Financial capital is obtained through fiscal resources that similarly allow stakeholders to access a broader range of resources. Infrastructure, usually in the form of building, housing, and physical infrastructure assets, is the built capital that is used to support community stakeholders. Lastly, natural capital refers to the physical features of a particular location, including weather, isolation, and natural resources that provide raw materials for production (Cochrane, 2006).

Under a multiple capitals framework, the iterative and recursive relationship among these capitals are emphasized. Important in the community development process is that gains in any of these capitals could 'spiral up' and lead to gains in other forms of capital. In turn, this spiralling up helps to build local capacity.

Local government

The second section of this chapter turns to the topic of local government or local administration. Depending upon the national context, the specific jurisdictional language and structure will present differences. For our interests, we shall use the generic concept of 'local government' to refer to the mechanisms by which place-based communities in resource-dependent regions look after matters of local responsibility. In considering the role of local government as an institutional actor in the restructuring of rural and small town places, our principal message is that, while these local government institutions are encountering a time of dramatic change, they exist in a very confusing context of transformation where the redefinition and re-positioning of the role and activities of local government seems so important, but where they continue to be hamstrung by outdated structures and legislative contexts that limit the effectiveness of their responses. Before we get into specifics respecting the challenges facing local governments, and their response to those challenges, we set out a number of general observations.

As has been noted in the literature, local government is perhaps the most important, and yet most neglected, level of government (Douglas, 2005). Its importance is partially the result of its 'immediateness'. It is both accessible to local residents and it provides many of the most basic and needed services for those residents. It is also involved directly in local processes such as regulation and planning that protects and governs the most significant investments people will make – property and housing. Depending upon the jurisdiction, across OECD countries some local governments also have responsibility for social and welfare service provision. Local governments also inherit the mantle of 'local voice' for representing the concerns and aspirations of residents. Given these immediate and important roles, concern about neglect in studying the changing role of local governments builds from the observation that it receives far less attention than other forms of political, policy, or governance research. As we shall see, while the impacts of Neoliberalism and the globalization of the economy have dramatically changed the role and actions of capital, labour, and states, local government has been relatively unchanged in terms of the fiscal and legal tools at its disposal. As a result, it is often ill equipped for the challenges it now faces, the tasks now being asked of it, and the more general world in which it must function.

Historical context

There has been a range of long-standing structural impediments to the institution of local government. One of the first of these involves the small size of the local community and thus the small size of the local government itself. Historically, this has limited the capacity of local government offices in rural and small town places to undertake various forms of activities. Again, depending upon the country, this may not always be the case. In some states, such as Japan and Finland, state supports and fiscal transfer structures allow local government offices to be relatively larger and more fully staffed than in places such as Canada and the United States.

Coincidence with the smaller size of local government in rural and small town places is the limited fiscal capacity of those local governments. In some jurisdictions, local governments are dependent upon a local property tax for most of their operating revenues. In this case, the jurisdiction is limited by the number or value of local property development. In other jurisdictions, local governments are more dependent upon transfers from senior governments that are calculated on a per capita basis. In this case, it is the number of people residing in the local jurisdiction that governs the size of the fiscal transfer. These differences in sources of fiscal revenue for local governments can dictate the focal points for local government activities with respect to community development and transition planning.

Another historical context of importance for local government is their 'location' in the governing hierarchy. Depending upon the constitutional structure of states, local governments may or may not have protected rights and responsibilities. In Canada, the Canadian constitution recognizes only the federal and provincial levels of government. Local government is considered a responsibility of provincial governments and thus exists at its pleasure. Local governments in this context operate under the legislative and jurisdictional framework that provincial governments set for it, and these frameworks can be quickly changed, depending upon the desires of that senior government. By comparison, in New Zealand, the central state through parliament controls the powers provided to local governments (Asquith, 2012). Under this system, Neoliberal reforms have been equally targeted to reduce the role of the state, while placing considerable more responsibilities on local governments as they were provided with the power of general competence that gave them "full capacity to carry on or undertake any activity or business, do any act, or enter into any transaction" with "full rights, powers, and privileges to do so" (Hide, 2011, p. 7).

Another more universal challenge for local governments has to do with questions of boundaries. First, there are the jurisdictional boundaries of the local government unit itself. These boundaries can create structural impediments to regional cooperation and to more holistic and integrated community, land use, and economic development planning. Historically, many local government boundaries tended to be tightly constrained around the built-up portions of a local area; although, through the late 20th century in some resource-development regions, these jurisdictional boundaries were expanded in order to provide local governing structures over rural territories which might be subject to intense development pressures. A second aspect to the jurisdictional boundaries question includes the limits to the management, planning, and regulatory activities that a local government can undertake.

Big industries and small local governments

The earlier introduced challenge of small communities with very large industrial interests operating within their territory is an opportunity and a challenge for local government. As noted in Halseth et al. (2017a), local governments in this context have historically tended to leave management of the local economy to the large industrial players.

Under the pressures of social, economic, and political restructuring, there can be a powerful and lingering nostalgia for the 'paternalistic' firm. Halseth (2005) describes how many local governments in resource-dependent places will talk about economic diversification, but how they will often focus their attention on attracting replacement large industries. Workers and local government leaders remember somewhat nostalgically earlier times when a single large employer meant relative stability and good paying jobs.

As a consequence, local economic development often focuses upon the 'big win'. That is, trying to bring another very large employer to the community. This is an attractive strategy given that large numbers of direct jobs and local spending will create a multiplier effect across the entire economy. This is also the familiar pathway for resource-dependent communities that had grown used to a single large local industrial employer. Even where the economic development plans call for economic diversification, the desire to secure a replacement single large industrial employer often overwhelms efforts towards meaningful diversification.

One of the topics which brings together local governments and industrial players concerns that of 'corporate social responsibility'. As Neo-liberal policy transformation has reduced the role of the state in supporting resource-based communities, and enabled capital to be more mobile and flexible, there have been concurrent arguments about the role of industries, their social license to operate, and their social responsibilities in the localities where they operate. Historically, resource companies played a significant role in the reproduction of both labour and communities. In the post-war Keynesian public policy framework, these roles were increasingly assumed by the state, in part, to reduce the burden on capital and shift public sector expenditures to taxpayers.

As noted by Cheshire (2010), some private-sector actors have become re-engaged with the resource-dependent communities within which they work, mobilizing partnerships between public, private, and voluntary sector interests as part of a new form of local and regional governance. This participation also fits with the corporate need to demonstrate social responsibility and thus achieve a sort of 'social license' to operate (Mayes, 2015). Both social license and corporate social responsibility are flexible terms, however, and often lack any form of regulatory framework – they are thus subject to interpretation and change. In their study of corporate social responsibility in the British Columbia community of Logan Lake, McAllister et al. (2014) argue that a deeper exploration of the concept of 'corporate citizenship' may assist with evaluations of the roles, responsibilities, and actions of industrial capital in resource-dependent places. They also argue that the transformations in resource industry activity, especially including long-distance labour commuting and distributed servicing may make the local scale of study less suited to topics such as corporate social responsibility. They suggest instead that "effective 'corporate citizenship' practices might be better realized through participation in a regional 'place-based governance' strategy along with rural and remote communities that have often been marginalized by dominant political and economic interests" (McAllister et al., 2014, p. 312).

Contradictions

To begin to explore the challenges facing local government, we need to understand the contradictions now embodied within local government. A first contradiction concerns the relationship between representation and power. On the one hand, local government is, generally, directly elected by the citizens of the community. While there are differences between countries in terms of eligibility to vote, it is a direct form of democracy over the management of local affairs. On the other hand, local government can only function within the authority and powers granted to it by non-local bodies. There are limits to this jurisdiction and there are even exceptions – such as the common exception, whereby the agencies and actions of senior governments are not bound by local government regulation. In some countries where local governments have an effective veto over large industrial or resource-development proposals within their territory, senior levels of government have changed approval processes to override these powers based on vague arguments of developments being 'in the national interest'.

A second contradiction has grown out of both the recurrent crises of capital and the policy transformation from Keynesianism to Neoliberalism. Local governments have limited local revenue sources – whether derived from local sources such as the taxation of property or derived from transfers from the central government. Consequently, the roles of local government were historically quite limited. The political science literature on local government talks about its role in the delivery of basic infrastructure services such as roads, sewer, and water supply, as well as the regulation of land uses to help manage those services and expenses. In other cases, these are more expansive and include social and welfare services. Depending upon the context, local governments have also been where property interests or local business interests have tended to be over-represented in political decision making and participation.

The political science literature also talks about post-war local government policy debates about breaking away from simply providing services to property versus broadening to provide services to people such as social welfare, parks and recreation, and the like. Citizens pushed for improved basic infrastructure and greater services to people because local government was accessible to them – it provided a convenient local setting for advocacy by different interests. It also provided an accessible electoral process, whereby representatives could rightly fear for their political survival if they were not mindful of populist initiatives. Against such pushes, the limited fiscal capacity of local government and its very real infrastructure maintenance and management costs, did tend to damper forays into other policy or service areas.

An important consequence of the general Neoliberal policy shift has been the need for local government to take more of an entrepreneurial stance in its community and economic development planning. Hayter (2000, p. 288) describes this transition from 'managerialism' to one where resource-dependent towns "had to contemplate becoming entrepreneurial and try to create ideas for development from within the community". He uses the notion of an 'unruly' process to describe this new reality for local development, because it was so different from the organized and structured ways the community had worked under the earlier Fordist management framework of large corporations and senior levels of government.

The shift to an entrepreneurial approach on the part of local government is also driven by the downloading and offloading actions of senior governments and their ethos of 'enabling' local communities to divine more of their own economic and community development future. As noted by Tennberg et al. (2014, p. 6) in the Barents region, the paradoxes of Neoliberal governance "also compel local communities and individuals to tackle an increasing number of new responsibilities to ensure the economic and social well-being of local communities". The shift, however, has a number of structural challenges. One of the first key challenges has to do with timeframes: economic diversification generally takes a great deal of time, while the needs of families displaced by resource industry restructuring are immediate.

Another challenge is that senior government has reduced or eliminated many of the key supports (in terms of services and institutional connections) that local governments need to realize new development strategies. This is exacerbated by both relative and absolute distance. Senior governments tend to be 'far away' from rural and small town resource hinterland regions and their urban orientation makes it difficult for them to understand the needs of resource regions, despite the fact that urban population and governments are still very dependent on the export wealth generated by resource commodities from non-metropolitan areas.

As part of their own response to restructuring, large industries have also stepped away from their historic economic stewardship role within resource communities. As described in earlier chapters, they are under intense pressures from increasing competition, long-term relative price reductions in commodities, and shareholder expectations of returns on investments. The result has been reductions in the fiscal and leadership roles that industry used to play as they stepped back.

A first response to both the transition from managerialism to entrepreneurialism and the deep restructuring of resource industries has been for small communities to adopt a local economic development approach. This approach, with deep roots in local boosterism traditions, seeks to entice new economic activity into the community. Towns will set up economic development offices to do this function and will use tools such as property tax relief and other such incentives to lure in new businesses. The 1980s, in particular, witnessed a significant uptick in the addition of 'economic development' functions to many local governments.

Such efforts have, however, met with limited success, especially limited success in producing transformative change. Markey et al. (2012) examined the process of doing local economic development planning and identified critical problems with its execution. Building on a three-stage model, they noted that effective local economic development involved a strategy stage where local participation and understanding could be marshalled into a broad process of visioning and goal setting – all linked to local values and assets. A second stage included looking at 'blue-sky' options to identify potential opportunities and then assessing the business case for those opportunities against the context of the place (its assets, infrastructure, regional setting, etc.). The third stage of the generic local development model involved active implementation, together with monitoring and the flexibility to respond to changing circumstances and opportunities over time.

The critique is that local economic development planning has been 'stuck in the middle' (Markey et al., 2008b). Too often, local or senior governments would fund a blue-sky study of economic possibilities for a region or community undergoing economic distress. When looked at collectively, these plans almost always have a similar list of options. There is little or no attention to stage one (local involvement, participation, understanding, vision, and assessment of the context of place), and the funding almost never includes support for effective implementation or the monitoring / adjustment that is needed over the 10–20-year periods that research has shown is needed to transform rural and small town economies. In the end, the local economic development approach has seen lots of action, and consumed lots of fiscal resources, but produced little.

A final key challenge has involved the need for small places to 'scale up' to work together at the regional level. There is an inherent synergy between local and regional actions and activities around social and economic development (Porter, 2004). This synergy demands awareness and cooperation at both local and regional levels. It also needs institutions or forums to support dialogue and awareness building. As rural and small town places are continuously challenged by capacity issues, working more cooperatively as a region allows communities to share workloads and build on collective skills and expertise. At the same time, the region remains a reasonable scale for understanding impacts and working towards solutions. The region is also a broader and more appropriate scale at which to evaluate the large infrastructure investments needed to assist both regions and places in constructing a more competitive framework for attracting economic activity and people.

Local government and community development

Crises of capital put at least four direct forms of pressure on local government with respect to its capacity to undertake community development work. The first is that as local resource industries downsized or closed, there was a reduction in the economic and taxation contribution these industries made to local government (see also Chapter 7). This reduced the fiscal capacity of local government to do even its most basic tasks. Second, local layoffs from such downsizing or closures put pressure on local governments to provide relief services and to expand its efforts to attract new jobs and economic activity to the community.

A third implication of resource industry restructuring focuses on the reduced contributions they made to state revenue. Relative to historic shares, these fell as resource industries either closed or appealed to senior governments for relief from taxation and regulatory burdens. As noted, the reduced state revenue was, in turn, often passed along to local governments via the withdrawal of state services, reduction in transfers, or reduced participation in jointly funded infrastructure projects. A fourth is that, commensurate with reduced state revenue, central governments began to download more and more functions and responsibilities to local governments using the disingenuous Neoliberal argument that local government could now 'exercise choice' in whether they wished to adopt those functions, services, or activities. Of course, the problematic aspect of this action by central states is that while they downloaded functions / responsibilities, they failed at the same time to download additional fiscal powers or resources to local government and they did not download any additional jurisdictional authority. For DeFilippis and Saegert (2012b, p. 380), "While the scale of organization of capital and labor has been globalizing, the scale of government as it relates to community development has been shrinking". Now charged with increasing responsibilities for community development, local governments do so with the constrained fiscal resources and jurisdictional authority of the Fordist industrial era that are now outdated.

Within this contradiction of an increasingly globalized economy and an increasingly localized base for devising options and responses, local government has little choice but to take up community and economic development functions. In part, this fits with earlier local government advocacy, when communities grew tired of being dictated to by the centralized decision making of the state in terms of economic and resource-development policy. Instead, local government argued that they needed to be involved and to play a greater role. For local governments, this was a twofold argument. They argued that they knew the aspirations of citizens and thus could better match the type of economic activity with what locals wished to see (thereby also reducing resource conflicts). They also argued that they wanted increased involvement and control to ensure that more of the resource wealth generated locally would stay in the community and region. Under fiscal pressure, and via the Neoliberal policy turn, the state generally acceded to the first argument but refused to budge on the second.

The placement of more community development responsibility with local government has been accompanied by the emergent language of Neoliberal approaches – namely, the default to markets and market mechanisms. Such a reliance on market mechanisms, however, presents some significant problems for community development generally. For example, "All theories of community development recognize that markets as they exist are not the ideal free markets of economic theory but instead products of the existing distribution of power and resources" (DeFilippis and Saegert, 2012b, p. 378).

As noted in the critique of Neoliberalism, this use of an approach with inherent self-contradictions creates challenges. Building from an argument that community development needs to redress some of the more distinct distribution problems within the market, DeFilippis and Saegert (2012b, p. 377) highlight that:

> Since 1980, pretty much all efforts to resolve problems of social equity rely on market mechanisms ... Reliance on market mechanisms to resolve equity issues often falter because the aims and goals of successful markets and equity are not the same. Communities in need of development are by definition failing in the market. The goals of community development are inherently redistributive.

In response to challenges related to new pressures for economic and community planning as well as continuing limited fiscal constraints, local governments have experimented with spinning off or downloading various activities and functions. One common trend has been for local economic development activity to be removed from local government offices and placed with a local development corporation. This arm's-length organization reflects an ideological approach to working with business interests. Interestingly, examples from around the world show that if an external economic development corporation had been created in the past, the trend might be to bring that activity back within local government purview, while in other cases, where that activity had remained within local government, there are pressures to transform it into a stand-alone economic development corporation. Dissatisfaction with the ineffectiveness of local economic development efforts and mechanisms is generally behind these 'flip-flops' in approaches. This also makes clear that resource-dependent rural and small town places need a different approach, one that builds on local assets and aspirations, is embedded within both local and regional development planning frameworks, and which have extensive support (policy and fiscal) from senior governments. Until there are such wider changes in approach, locally isolated economic development approaches will struggle with limited success.

In contrast to the smokestack-chasing approach, writers are pushing for more contextualized approaches that recognize not only the fluidity and pace of the global economy, but also the need to seek appropriately scaled, nimble, and globally competitive opportunities. The international collection of rural development chapters in Halseth et al. (2010a) chart the rise of a place-based approach to community development that is both holistic in bridging the political, the economic, the social, and the cultural, and is nuanced in recognizing the uniqueness and 'situatedness' of different rural places and regions.

The importance of recognizing uniqueness and situatedness, together with the reimagining and re-bundling of local assets and aspirations, has been translated poorly in many instances.

Woods (2010) addresses the problems that arise from a near 'faddish' adoption of place-based development approaches. Beginning with the argument that space simply "cannot be discarded in an analysis of the next world economies" (Woods, 2010, p. 168), he argues that globalization has created a framework whereby attributes of place within the structural constraints of space are now more highly complex and interwoven than at any other time.

Local government – change and readiness

Uneven development, the uneven impacts of restructuring, downloading, and offloading by senior government, and the uneven application of Neoliberal policy shifts in responsibility for community sustainability to increasingly lower levels all put pressure on local governments. These pressures collate around the need to be 'ready' for sudden changes within the global economy and their local impacts. They also collate around the need to be ready within the context of the relatively scarce human, fiscal, and capacity resources available to local governments to support such planning.

To explore community responses to social and economic restructuring, there needs to be a greater understanding about how to achieve change. There is a considerable body of work that examines different models of the organizational change process (Armenakis and Bedeian, 1999; Kettner et al., 1985). In very general terms, the change process can be thought of as involving some generic 'stages'. First, there is the need to identify the opportunity for change, including identifying the urgency, the initiators, the potential benefits, the focal points, and the people likely to be involved in planning and implementing change. Moving forward, there is a need to clearly define the problem and analyse appropriate information to inform decisions about change. Goals and objectives need to be strategically designed to facilitate the transition into action. Next, attention is needed in designing and allocating resources to implement the change effort through appropriate policies, programs, or projects. Momentum for sustained change is nurtured through promoting short-term successes. The monitoring of change helps to ensure that activities are completed and necessary adjustments are made. Evaluation then provides an overall review of the process and information to make the change effective and efficient.

Within the organizational change literature, researchers have described several topics within new institutionalism debates that are related to program design, structures, conflict and culture, learning, capacity, and barriers to change. These institutional topics recognize problems around the limited recognition of the need for change; a lack of clarity around roles, responsibilities, or jurisdiction to resolve the problem; limited information, resources, and training; inadequate leadership or human resources capacity; limited credibility by leading stakeholders; inadequate financial resources; denial or resistance to change; and underdeveloped structures to guide change (Battilana and Casciaro, 2012; Weaver, 2014).

Change and resource-dependent small places

With a focus upon sustainable development, Nozick (1999) articulated a vision for remote, rural, and small town communities that concentrated upon local assets and capacities. Plugging economic leakage and securing more of the local resource base into local hands are touchstones of this articulated vision. In terms of local 'ownership' of the various community economic development resources, Nozick (1999, p. 10) argues that, "Gaining community control and decision-making power over the allocation and use of local resources – land, capital, industry, and human resources (delivery of community services) – is essential to building sustainable communities". Local control, and the multiple interconnections of local economic activity this can create, provides the potential recirculation of wealth needed to grow the local economy.

Also beginning with an interest in sustainable development, Bryant (1999) addresses the unevenness that has resulted as local governments have increasingly assumed responsibility for managing change. In addressing how communities experiencing dramatic restructuring can prepare and purposefully engage with the processes of change, Bryant (1999) poses a series of questions. These include:

1   Is there a long-term planning process in place that is continuous, holistic, and strategic?
2   Is there real community participation in the processes of community development planning and implementation and adequate representation of, or communication with, the various orientations in the community?
3   Are there effective links between the public, private, and voluntary (or 'third') sectors in and managing change?
4   To what extent are the values and interests of nonresidents taken into account in the process of planned change?
5   To what extent are economic, social, and environmental values integrated into the development planning process and the various projects and initiatives being implemented? (Bryant, 1999, pp. 80–83)

Neoliberal limits to local governments

For local governments in resource-dependent regions, there are numbers of limits or constraints that have been exacerbated by the continued implementation of a Neoliberal policy approach. In concert with having more 'freedom' to act on local matters, local governments are also burdened with increasing responsibilities and expectations (from both senior governments and from local residents and businesses). Coincident with the offloading or outsourcing trends noted in previous paragraphs, these pressures have led to the expanded use of specialist organizations, larger jurisdictional partners, or consulting firms to address community and economic planning efforts. While there are clear benefits with accessing external expertise to address particular problems (especially for smaller local government offices), the implementation of such activities can sometimes undermine local capacity. In other cases, the resulting 'products' may not be as attuned to local circumstances and contexts as they need to be in order to effectively support community or economic development. In still other cases, there may not be the commitment to local participation and involvement that is so critical to building the foundations for effective implementation.

Another topic that has emerged as important in communities experiencing new industrial economic investment and activity is competition for staff. Resource industries are generally able to pay a more competitive salary and benefits package than are local governments or community organizations. This loss of staff can mean that local governments struggle with diminished capacity at the very moment when they need to be at their most effective in managing the upswing in local activity and investment that impacts everything in the community from housing and infrastructure, to services, public safety, and social welfare. Over the longer term, continued competition for staff means that local governments will have challenges finding and retaining needed replacement staff. These pressures are exacerbated in Aboriginal communities, where staff resources are also overwhelmed to respond to the number and breadth of requests they receive as industries and governments adhere to their duty to consult in territories that have outstanding land claims or native titles.

Coincident with the longer-term economic upswings and downturns associated with resource commodity demand and price cycles, local governments must now contend with the accelerated pace of industrial change. This pace of change can be very dramatic at the local level, and yet local governments are just not able to react with the same speed. They may need to contact community partners or senior government partners in order to put together plans on how they might respond. Such takes time and is nowhere near as rapid as industrial decision making.

Within the context of these challenges, local governments are also struggling with outdated policies and procedures. In some communities, this is the result of neglect where a local or regional resource industry may have been functioning continuously for a long period and, despite ongoing restructuring that likely reduced jobs and local benefits, the community and local government had idly watched the contraction. When dramatic changes such as industrial closure do come, these places are not equipped with the up-to-date policies, plans, or procedures needed for respond effectively. In other communities, where there may be the additional stresses of multiple economic upswings or downturns across different sectors, the time often cannot be found within small local government offices for updating critical policies and procedures.

Further exacerbating the challenges facing local governments have been the changes in mindset around investments and expenses that come with Neoliberal initiatives to reduce the tax and other burdens on local industry. A key result has been that local governments have been lagging in terms of their investments in infrastructure – most especially, the infrastructure for modern governance and government management through technology. Information management systems have struggled to keep pace with the rise of the social media world, and in other cases, local governments struggling under fiscal constraints have actually reduced or downsized the emphasis they put on information management. The same can be said for a range of service deliveries as managed by local governments.

When challenged, the response of local government is important. As noted in Chapter 6, senior governments have frequently adopted an almost 'neo-Keynesian' approach during recessions by injecting stimulus funding for infrastructure; although projects have not always aligned with local priorities, and have often challenged local capacity with tight funding deadlines. As one marker of this more 'reactionary' era in public policy approaches (see Chapter 6), we find that these programs, however, were often subject to change or cancellation. Local governments have responded with "a great deal of creativity and maturity in their engagement with senior government programmes and priority funding areas" by refusing to blindly pursue any or all funding opportunities, many of which do not complement, but can even threaten, community development under conditions where budgets and resources in other priority areas may be mistakenly restructured to chase any opportunity presented (Manson et al., 2016, p. 109). Local governments are also becoming more strategic in the resources they pursue by focusing on developing and retaining local assets rather than chasing senior government grants that often do not reflect local pressures and needs. Nurturing working relationships across local and regional business stakeholders is becoming more important than focusing too many resources on business attraction.

Local governments are challenged at this very confusing time of transformation. As Manson et al. (2016, pp. 105–106) write:

> Historically, the local response to boom periods involves an abandonment of coordinated planning in favour of disorganised opportunism in order to capitalise on an economic upswing. During bust periods, planning is again abandoned in order to focus on assistance to industry to maintain local operations, and to labour in order to mitigate the loss of employment. Local governments have traditionally responded to busts by slashing services in an effort to 'ride out' the downturn. Especially vulnerable to cuts are community development and local quality-of-life services.

They are challenged in redefining their role and also challenged in redefining the scale at which they should work. To this latter point, the following section deals with the issue of scaling-up local government strategies and activities to a regional level.

New regionalism

In the post-war Keynesian public policy framework, regional development through active state intervention was common across developed economies. Indeed, many rural and small town resource-dependent communities in more remote areas thrived under this policy framework. However, as regional economic development policy and planning fell out of favour through the 1970s and 1980s, it dissolved under the transition to a more Neoliberal public policy approach and the early years of a more intensely globalized economy. For Tonts and Haslam-Mackenzie (2005. p. 183), Neoliberal reforms are part of a "political strategy based on deregulation of the economy, privatization, a reduced commitment to social welfare, and a focus on international competitiveness". It places increasing emphasis on downsizing the participation of the state and places greater emphasis on individuals, the non-profit sector, and the private sector in delivering services and managing economic transformation. But as a consequence of the failures of this policy approach to support rural and small town transition and competitiveness, there has been over the past 20 years a resurgence of interest in 'the region' as a framework for economic renewal in lagging rural and resource-dependent regions.

Research interest into the re-emergence of place as an important organizing construct for understanding how processes of social, economic, and political restructuring unfold has led to a parallel re-emergence of an interest in the region. Markey et al. (2008a, p. 410) ague that:

> This contextual turn is found in a variety of ongoing rural research themes, including post-productivism, conceptualizations of the role of competitiveness within the new economy, and the adoption of a territorial, rather than sector-based, orientation to rural policy development. Each of these themes provides insight into the role and meaning of place within the rural development process.

One of the reasons for the attractiveness of a regional analysis is that the region represents a manageable scale for not only understanding multiple impacts, but also for designing mitigation strategies.

Another critical reason for a renewed interest in regional cooperation is that in rural regions requiring significant new infrastructure investments, it is only by working at a regional level that appropriate and transformative regional infrastructure investments can be made. In a competitive environment for senior government support of infrastructure projects, it is only by working together that large numbers of small places can gain the political capital required to win over such investment decision making.

New regionalism also provides one opportunity for regions to resist the 'one size fits all' policy approaches of many senior governments. It allows for the exercise of what Massey (1984b) had recognized: that places and regions are unique in that there are combinations of assets, populations, histories, and circumstances accrued over time that create unique places and regions.

For writers such as Markey et al. (2012), new regionalism includes the mechanical process by which small places 'scale-up' to work more effectively as regions in both the public policy arena and the economy. At the state level, the regional construct also supports new policy aimed at addressing rural development challenges. Drawing on models from elsewhere, Ireland set in place a structure and organizing framework under a Commission for the Economic Development of Rural Areas (CEDRA, n.d.). Under the direction of NORDREGIO (2011), broad attention to the supports needed for rural innovation and development are under way for the Nordic countries. Norway has specific regional programming, for example, around innovation, research, and development to support a break from 'path dependency' in many smaller and more remote regions (Jakobsen et al., 2012). In Finland, the concepts of regions and clusters have been brought together to provide a spatial framework for investment in innovation (Makkonen and Inkinen, 2014). But innovation requires attention to the local / regional context. For Shearmur and Bonnet (2011, p. 266), this means that innovation works best for areas that "are sufficiently large and diversified to exploit their own innovations and capture, locally, the benefits". As noted in an OECD policy review (Jean, 2014, p. 9), this attention to capacity as well as assets is critical: "Each rural region is different and has its own set of opportunities and constraints. Local actors must come together to build an understanding of how best to use these resources".

One of the reasons for renewed interest in a territorial approach is a rejection of older comparative advantage economic development approaches for more competitive advantage economic development approaches. Markey et al. (2006, p. 25) argue that,

> regional competitiveness is multi-dimensional, mixing traditional factors of infrastructure with more ethereal factors such as amenities and social capital. Interestingly, this multidimensionality is reflective of new conceptualizations of the region itself. If the 'old' region was previously seen as a subordinate administrative unit, the 'new' region represents a territorial hub for production, planning, trade, education, and innovation.

But regional competitiveness needs a supportive institutional environment for coordinating strategic intervention and for advocating beyond individual jurisdictional boundaries. This institutional density or 'institutional thickness' of supportive institutional environments is vitally important to the successful development of regional competitiveness.

A key part of institutional thickness is the degree to which organizations can work together. This involves not just local government, but also other formal and informal groups and organizations. As such, within new regionalism research and writing, attention moves beyond questions of government to include questions of governance. As Markey et al. (2008a, p. 411) write:

> governance regimes are prominent within the place economy. While potentially stressing local capacity, there are two place oriented byproducts associated with this transition. First, governance implies a re-drawing of the lines of accountability and control, away from centralized state power, to be dispersed amongst a greater diversity of local and extra-local actors

and institutions. As part of this re-mapping process, governance mechanisms may initiate regional dialogue and cooperation, altering the directionality of traditional heartland-hinterland flows of communication and resources. Second, the participation inherent in governance fosters a sense of ownership, over decisions and ultimately resources, that may not have existed under previous top-down regimes. Thus, place not only reveals a greater variety of assets, it may also instill a sense that those assets are local and may be used for local purposes.

Through new regionalism, regional governance arrangements are playing increasingly important roles in decision making and community development.

A final item under the imperative for smaller places to scale-up their representation links to the previous chapter's interest in senior governments. It is difficult for rural and small town places to find an effective voice when dealing with senior government as they are often far from the centres of decision-making power. They are also out of the media spotlight (except when there is a catastrophic natural or economic event). One option now being tried in many parts of the European Union is the use of 'rural parliaments' (Woolvin et al., 2012). Rural parliaments provide a venue through which rural and remote places and regions can come together in policy dialogue and debate. They also provide the opportunity to craft policy and program proposals, to share experiences that demonstrate how they are all being buffeted by restructuring forces, and to raise awareness about non-metropolitan issues.

Conclusion

The political economy of local government institutions is wrought with contradictions and these limit its effectiveness in leading change in this period of uncertainty and unevenness. In resource-dependent rural and small town places, social, economic, and political restructuring has resulted in challenges to the understanding and operation of local government. Local elites in business, labour, and industry are often deeply wedded to the old resource commodity economy and resistant to change. The strong senior government policies and supports that built these resource hinterlands have been undermined and reduced as part of the Neoliberal public policy transition. As local government takes up the opportunity to plan for its own economic future, it is pulled by its own constituency and hampered by its lack of control over key policy and fiscal levers to mobilize most forms of any agreed upon local strategy.

And yet, resource-dependent rural and small town places remain the focal point for decisions and activities by international capital and senior governments. They are rooted in places and they experience the dramatic consequences of contemporary political, economic, demographic, and environmental change. This chapter has defined the concept of place-based communities and why understanding the institutional structure of those communities, specifically through the mechanism of local government, is so important for equipping those places to become more resilient over time. The chapter concluded with a discussion of new regionalism and the imperative for small places to be able to work independently on strategies important at a place-based scale, while also being comfortable working together with their neighbours and a wider array of partners at a scale important for decisions that affect regional competitiveness and the capacity to undertake renewal.

10 Civil society

Introduction

The roles and pressures of civil society have been shaped by the underlying motivations and strategies pursued by both industry and senior levels of government. As we continue our consideration of the implications of social, political, and economic restructuring, we introduce the topic of local services as they have been organized and delivered through public, private, and voluntary sectors. Following this, there is an introduction to the concept of civil society before exploring the voluntary sector specifically, providing a background of definitions and a sense of the scope and scale of voluntary-sector activities. This then provides a foundation to discuss how evolving staples-dependent economies in resource regions were supported by, and also profoundly impacted, civil society groups. In this context, we focus on the role of civil society, as it was supported by Keynesian policy and programs, as it helped the expansion of the staples economy in resource hinterlands. We then turn our attention to the early stages of transition to Neoliberal public policies that began to undermine the capacity and renewal processes of resource-dependent regions. As industrial restructuring continued, we discuss the implications of labour-shedding technologies and new mobile labour arrangements on civil society structures in resource-dependent regions. This discussion also highlights how other trends such as demographic aging are impacting the infrastructure and operations of civil society groups. The chapter finishes with a discussion of some of the responses voluntary and non-profit sector organizations have undertaken as they seek to adjust to the increasing pressures of social, political, and economic change, and adjust to the opportunities that new information technology and other tools provide.

Services

As noted earlier, the delivery of services to residents in remote, rural, and small town resource-dependent places has become increasingly problematic. To start, smaller places and rural regions face some very traditional geographic challenges around the fact that large distances and low population densities typically make the delivery of services more costly on a per capita basis. While examples of the application of new communications technologies can be found, and do seem to have a positive impact on reducing rural service costs and increasing rural service delivery, they have not been as widely adopted as one might have expected by this date. These traditional challenges have been exacerbated by senior governments who have chosen to either close rural and small town services, to reduce funding to support those services, or to offload responsibility for them to local governments or the local voluntary sector.

If we look back in time, it is possible to identify three general eras of services provision that have transformed rural areas (Halseth et al., 2003). These have impacted both the stability of the local services base for rural communities, and the level of involvement or responsibility that residents have in delivering local services (Grafton et al., 2004). Prior to the Second World War, rural and small town places in developed economies were generally self-sufficient and isolated, with limited service provision or coordination by either private or public sectors. Communities defined what was most important, and core local services were tailored to the needs of the community. This laissez-faire policy approach meant that rich communities were 'services rich' (i.e. with well-developed recreation centres and health centres), while poor communities had relatively fewer services. The economic boom following the Second World War was accompanied by Keynesian public policy approaches that expanded the state's role in developing a stronger social safety net with more investments in education, health, employment, and other social programs (Blake, 2003).

Keynesian public policies supported trends linked to the community development and community economic development needs of resource-dependent places. Policies and programs addressed a wide sweep of social, economic, and political matters including employment opportunities, income levels, and the general standard of living and quality-of-life matters. For Sheppard (2009, p. 548), the more progressive Keynesian approach focused on "continual state intervention to manage the contradictions of capitalism to the benefit of the nation and its least well-off citizens". This extension of services to support the industrial resource-development policies of the post-war era not only followed Keynesian policy logic, but they supported the economic investments and workforce needs of capital. More pointedly, an investment in public services acted to support industrial investments by providing the infrastructure and services that workers and their families would need. In turn, this assisted capital by reducing their costs for reproducing labour and similarly reduced their costs from labour turnover.

In the early 1980s, a third era of services provision was initiated. The adoption of Neoliberal policy approaches prompted a retrenchment of public service delivery with more responsibilities being downloaded to local governments and contractors in the private and voluntary sectors (Tonts and Haslam-McKenzie, 2005). This has left remote, rural, and resource-dependent communities with important challenges around services provision as large distances, small populations, and high operational costs make them less attractive for profit-motivated private-sector suppliers (Bock, 2016). In addition, non-profits in resource-dependent communities often lack the scale and capacity to take up increasingly complex public service contracts.

The closure of small offices and the regionalization of many services have been very difficult on small communities trying to create alternate development pathways in the globalized economy. Such retrenchment of services undercuts the capacity for small communities to respond to change and threatens short- and long-term community sustainability (Markey et al., 2010). Service and policy restructuring decisions, however, continue to be applied uniformly to rural regions, despite the diversity of needs and issues that exist across these places. In addition, the adoption of a Neoliberal policy approach has also moved the metrics for evaluating and delivering services from a focus on addressing needs to a focus on costs. Neoliberal public policy approaches have also pursued funding formulas that too often are biased to metropolitan contexts. As a result, resource-dependent regions have struggled to access adequate services funding, despite the fact that rural residents pay state taxes just like metropolitan residents.

Public services

When studying public sector services, we can generally distinguish between those services offered by senior versus those offered by local governments. While models differ between national jurisdictions, senior government public services often include those such as health care and education. Some of these, such as education, may have local organizing structures such as school districts or management boards, but the policies and budget allocations are typically tightly controlled by the senior government. They also include the local and regional offices of senior government ministries associated with such responsibilities as transportation, resource development, environmental protection, social services, housing, etc. Those public services offered by local government tend to be physical services related to property such as roads, sewers, and water supply, and the like. Local government also plays a key role in its community planning and local recreational services. Protective services, such as fire protection, policing, ambulance, and search and rescue are often provided by a mix of senior government and local government arrangements. In some jurisdictions, local governments also have extensive responsibilities with respect to social and welfare service provision.

Private-sector services

As noted, there are significant and long-standing challenges in the provision of private-sector services in rural and small town places. Distance and the lack of user densities make profitability difficult. As described in Chapter 7, small businesses are challenged by some critical operating constraints, including the health of the local and regional resource industry, the distance of the small community to larger shopping centres, retail spending leakage, and the availability of employees.

The public policy transition from Keynesianism to Neoliberalism has been accompanied by an expectation that government withdrawal from direct service provision would be backfilled by private-sector businesses taking up these new market opportunities. Private-sector contractors, for example, were expected to take up contracts for the delivery of transportation, employment, and health care services. In urban areas, where the density of clients and customers is high enough, such has been the case. In rural and small town places, however, many services previously provided by the state have not been continued.

Civil society

Existing outside of private and government spheres, civil society groups are independent organizations that can represent an array of common interests and needs that contribute to the well-being of the broader population. In general, they are mobilized as citizen actions through voluntary and non-profit organizations. As a coming-together of people outside the institutions of the state and the economy, civil society creates a context for social organization, participation, order, and action. As such, it is one of the theoretical and practical cornerstones of community development.

Central to civil society are collective associations of individuals, residents, or other actors (Massam, 1995). At the local scale, such collective associations are often of grassroots origins. At the non-local scale, such groups may exhibit a much more formalized and professionalized structure and organization. Participation in civil society is driven by many motivations. In addition to broadening social networks, volunteering helps people develop organizational skills and gain leadership training and career-related experiences. Through volunteerism, residents can strengthen their commitment to community and enhance the community's cohesion (Hanlon et al., 2014). It provides opportunities for diverse residents to get involved and make valuable contributions to improve the health, safety, and accessibility of the community, as well as to address service and infrastructure needs to improve local quality of life.

It has been noted, however, that across developed economies people are generally participating less in voluntary organizations (Putnam, 2000), as interest, demographics, and the advent of new information technologies shape and re-shape the way people associate. In resource-dependent rural and small town regions, population aging, household stresses from economic restructuring, increased workloads, labour mobility, and out-migration especially are all factors shaping lower levels of volunteer participation. Despite these challenges, the volunteer effort, as tracked through mean hours and median hours per volunteer, in some Canadian rural regions has increased (Sinha, 2015).

Voluntary sector services

One of the challenges in developing appropriate supportive strategies for the voluntary sector within resource-dependent regions is that many people have a relatively limited sense of voluntary-sector groups. They play a much wider and more important role than the stereotypical image of retired women running bake sales to support the purchase of new playground equipment. Their roles are much broader and they are often deeply embedded across a variety of community sectors, undertaking a wide range of activities.

Many definitions have been used to describe voluntary organizations. Among the more common characteristics are that voluntary groups are generally organized, non-governmental, non-profit, self-governing, and include people participating on a voluntary or unpaid basis (Barr et al., 2004). Other definitions describe how voluntary organizations serve a public benefit; how they depend upon volunteers for their governance; that they obtain financial support from individuals; and that they function with limited direct control by governments, other than in relation to grants or tax benefits (Marshall, 1999). These definitions can typically include organizations that may not qualify for charitable status, such as recreational associations, service clubs, and some forms of advocacy groups.

Within civil society, a range of names have been used to describe the more formally organized groups. Civil society organizations, non-profit or not-for-profit organizations, voluntary-sector organizations, and third-sector groups are some of the typical labels. Some of these, such as 'not-for-profit', may refer to specific legal requirements in jurisdictions where the funding and activities of the organization will not be subject to certain tax regulations, as long as they maintain that status. Usually there are clearly defined rules and policies that accompany such designations. Despite the range of labels in use, for the remainder of the chapter we will for simplicity sake use the term 'voluntary-sector organizations' or 'voluntary-sector groups'.

Notions of voluntarism typically exclude informal support obtained from family, friends, and neighbours, but can include acts of volunteering "by individuals whose occupations and professions are firmly embedded in the public and private sectors" (Hanlon et al., 2014, p. 133). Building upon these notions of voluntarism, a typology developed by Sullivan and Halseth (2004) can be used to understand different types and capacities of voluntary organizations. These would range from strictly voluntary (having no paid staff, no office space, and no government funding); to those with a mix of voluntary and paid participation (some paid staff and volunteers, office space, and a mix of funding sources); to those with a paid staff, office space, but which are still governed by a voluntary board of directors. Each of these different types can shape the capacity and resiliency of individual voluntary groups in various ways. Strictly voluntary groups, for example, can benefit from more informal structures and decision-making processes that can provide greater flexibility for quick action. Other types of voluntary groups, however, can benefit from more stable staff and office resources that bring the stability needed to obtain and sustain financial and human resource supports, as well as more formal governance structures, including boards, that can enhance their perceived accountability and legitimacy in order to obtain funding (Ryser and Halseth, 2016).

As noted, voluntary groups and activities go beyond stereotypical images and encompass a variety of definitional types and community sectors. There are local, national, and international services clubs that have engaged in fundraising efforts to support youth, housing investments, palliative care, community parks, and a range of other community needs. Committees may be formed to support senior government organizations, such as the delivery of health care in a community or region. At the local government level, it might include advisory committees or design panels as part of a local government approval process for development, or advisory committees that have a more standing role with respect to community planning functions.

There are also volunteer committees that support business and industry. One of the most recognizable and important of types of these are local chambers of commerce. Other examples might include voluntary boards supporting economic development infrastructure or, in some resource sectors, such as forestry, environmental certification programs, such as the Canadian Standards Association (CSA) and the Programme for the Endorsement of Forest Certification (PEFC) where voluntary community committees advise on impacts and monitoring of corporate activities. In other cases, such as mining in Finland, for example, there are standing voluntary committees set up that include industry and community representatives, whose task is to oversee the relationship between the two and mediate concerns.

Voluntary participation in groups is also important with a variety of civic infrastructure to maintain quality-of-life activities. For example, local museums and recreational facilities are often managed by a local voluntary board of directors. These boards typically have governance, as well as fundraising activities, as their key tasks. Voluntary-sector groups are also keenly involved as advocates for new facilities, teams, or assets such as recreation trails. Increasingly, one of the more important quality-of-life arenas where voluntary-sector groups participate is with respect to the organization and delivery of major events. These might be events that serve the local community, but they also might be larger events that the community would compete and win, such as a regional or national sporting activity. Not only do such events bring a profile to the community and region, but it also provides a significant economic stimulus.

Voluntary participation is also important in a variety of social and care service delivery areas. For instance, volunteer groups might organize to manage visits to elderly community members to make sure that they are well. Other groups might organize in order to achieve momentum towards a community good such as the development and operation of a seniors' housing facility. Others might address particular social needs, such as assisting with the organization of services for youth at risk. There are also, in any community, a wide range of voluntary-sector groups that are functioning as an ancillary component of ongoing services. One of the most common is the parent advisory committees that work with local schools and local school boards. There are similar advisory groups with organized child care facilities and the like.

Situating voluntary groups in resource towns after the Second World War

The post-Second World War period of community planning in remote, resource-dependent regions coincided with significant government and industrial investments to support the expansion of the resource hinterland (Markey et al., 2012). This expansion through planned towns was driven by a desire to recruit and retain more permanent residents, and thus improve the stability of the workforce for industry. The expansion of the staples economy and the resource frontier, however, was also extensively supported by civil society. Local residents provided services may have been formally or informally organized through churches, charitable groups, and nascent service clubs. Under a Keynesian public policy approach, voluntary-sector groups focused on particular local needs or issues, and they augmented core services provided by the state.

The planning and local government culture of these places was strongly shaped by notions of 'community', and various place-making processes helped to nurture social interaction. Infrastructure investments were strategically made to support the development of civil society groups and services. In many ways, resource frontier communities were amenity-rich. Either through their strong tax base or through industry donations, they were able to make investments in swimming pools, recreation centres, hockey arenas, curling rinks, golf courses, equestrian facilities, and more. To support this, industry leadership and workers were engaged in every facet of community life from being elected to town council to being on the boards of directors and volunteer positions with community services and voluntary organizations. Industry and small business members, for example, provided advice about investing money earned from donations or publications. Service clubs that drew extensively from local industry leaders and workers led successful initiatives to develop recreational facilities, community parks, playgrounds, campgrounds, hiking and ATV trails, and more. They were able to draw upon their industry connections to obtain both monetary and in-kind support that would range from donated use of industry equipment to trades and business expertise to support construction and operations of facilities and activities.

Civil society played an important role to strengthen the cohesion of industry's workforce and to strengthen community satisfaction and commitment to place. Many recreational teams for hockey, golf, curling, etc. were formed from industry workers on similar shifts. This shaped the relationships and support networks between families.

Emerging resource frontiers drew largely upon two important demographic groups that would influence the composition and activities of voluntary organizations. Until the early 1980s, resource towns often consisted mainly of young workers and their families who were recruited to support the industry workforce. As a result, everything from recreation programs to library programs targeted this young demographic. Even health care services were well equipped to address maternity needs and care for young families. Immigrant populations became an important second demographic force that shaped civil society. Aside from addressing labour shortages, immigrant populations brought diversity and cultural assets that would be drawn upon for everything from community dinners to fundraisers. As their engagement informed civil society programs and supports, they would also play an important role to recruit and retain new workers.

Implications of a mature staples-dependent economy and a Neoliberal policy environment on civil society

Pressures from industrial restructuring and Neoliberal policy transition have profoundly impacted civil society. Since the 1980s, resource industries have aggressively pursued efficiency through the adoption of labour-shedding technologies and the closure of smaller or less efficient facilities as a means to lower their costs of production and remain globally competitive (Edenhoffer and Hayter, 2013). The acceleration of these trends after the economic recession of the early 1980s reduced large shares of low-skilled jobs and increased the demand for skilled, professionalized, and specialized labour (O'Hagan and Cecil, 2007). This restructuring has also been accompanied by significant changes in labour allocation, including a replacement of permanent, full-time jobs with consultants, casual, short-term, or contracted labour (Holmlund and Storrie, 2002).

As noted above, policy changes aimed at reducing government expenditures have downsized, closed, or regionalized service supports (Halseth and Ryser, 2007). This has left unemployed or low-income residents in resource-dependent regions without adequate support to respond or cope with restructuring, develop new skills, or pursue new employment. In rapidly growing resource regions, where there has been significant expansion associated with new mining and oil and gas projects, there have been limited supports to assist those looking for work or to assist those who are struggling with the increased cost of living associated with rapid growth. As such, these economic and social restructuring processes have increased pressures and demands on voluntary-sector services. These organizations have had to expand their mandates to deliver programs around employment counselling, addiction, literacy, and other pre-employment programs. They have been responding to violence and abuse issues and assisting households to meet their basic nutrition and housing needs.

As industrial restructuring has increased demands for supports in resource-based places, however, successive rounds of employment losses have led to population decline or interrupted voluntary commitments in civil society as household pressures intensified (see also Chapter 11). Population decline affects not only the mean level of human capital within the community, but also the limited base upon which volunteer groups can draw members. This limits organizational capacity, and can also limit innovation and flexibility. Many voluntary organizations have struggled with declining membership and fewer active volunteers engaged to deliver services and programs. The voluntary landscape has also been changing with less interest in service clubs, declining church membership, and more interest in environmentally based or recreationally based groups. Increased competition for volunteers, burnout, high turnover among board and leadership positions, limited volunteer renewal, and a preference for short-term volunteer commitments are among the new realities affecting volunteer group capacity in resource-dependent places.

Community organizations in small places have long struggled with limited administrative capacity, staff, and volunteers. When dealing with the complex problems associated with restructuring, such challenges can be even more difficult (Johnsen et al., 2005; Poole et al., 2002). Over time, voluntary-sector groups may be able to draw upon, but not necessarily have continuous access to, financial or management skills. These groups may not have sufficient resources or training to address the range of issues associated with economic and social restructuring. Together, these can lead to different / conflicting approaches and unstable or inconsistent provision of services (Cloke et al., 2007). At the same time, most training and capacity-building supports are concentrated in metropolitan or urban settings and it is difficult for rural groups to access such supports. There is often a lack of skills to develop proposals suited to the increasingly complex world of grant applications (Simpson and Clifton, 2010), as well as a lack of time to create the onerous and lengthy applications or reports the now so often accompany program funding (these groups are after all busy delivering more and more services or supports). Where groups are busy delivering services, or chasing the dollars needed to keep those services going, there is often little time left for succession planning or job shadowing in order to help build and renew collective capacity. In addition, there is often little time left for the 'heavy lifting' work that goes into building collaborative partnerships within and across communities (Packer et al., 2002). In some regions, the voluntary and non-profit sector is not able to compete with the wages being made available in public and resource sectors and this can further degrade their capacity and limit organizational renewal (Ryser et al., 2014).

Voluntary organizations are also being sandwiched between state downloading of services, often with an expectation that the local private sector will play a role, and a private or industrial sector in many resource-dependent regions that is also under stress through recessions, increased competition, and retrenchment and consolidation, and thus not able to contribute, as it might have done historically (Peddle, 2011). Fewer options for funding have also been exacerbated by the retrenchment of industry and business support since the recession of 2008 (Ryser et al., 2012a). Poor commodity prices and delays in project development have meant that community donation programs are routinely postponed in accordance with boom-and-bust cycles. In resource-dependent regions, industries are reducing their donation programs; arguing that they are "paying substantial royalties to the government, companies publicly resist calls to provide infrastructure and services that they see as being the responsibility of the government" (Haslam McKenzie and Rowley, 2013, p. 376). This decline in financial support creates challenges for maintaining services and renewing both the mandates and structure of voluntary groups.

At the same time, there is also heightened competition for increasingly small pots of government funding (Graddy and Morgan, 2006). Government funding is also increasingly delivered on a short-term basis, thereby limiting the ability of voluntary groups to secure stable resources for staff (Walk et al., 2013). This creates a significant problem for those organizations as short-term funding does not provide adequate time for communities to mobilize and engage in larger community planning and consultation processes. Such short-term funding programs can interrupt the momentum for building relationships, planning, and mobilizing initiatives. With short-term funding programs, there is considerable uncertainty guiding the operations of these organizations from year to year (Hultberg et al., 2005). This not only impacts the ability of organizations to develop long-term management plans, but it also affects the ability of staff to develop stable support plans for their clients. One-year contracts, for example, only allow organizations to initiate new programs that build up community expectations that they may be unable to follow through on if the program funding is reduced or eliminated. All of these pressures targeting organizational changes take away valuable and limited resources from the delivery of services and supports.

Even though industry and policy environments have been changing, civil society has struggled to understand the extent and impacts of these transitions on their operations. Mandates can also impose challenges on voluntary groups in that they can limit the ability of these groups to recognize the changing pressures and needs in resource-dependent communities. Mandates can also be a challenge once a group fulfils mandate goals. Despite the need to renew mandates, organizations may seek to continue on perhaps for leadership or prestige reasons. As such, they may fail to renew their mandates, policies, capacities, and infrastructure in order to remain relevant and to re-position themselves in the community or region.

Lastly, resource-dependent rural and small town places have a very different context that affects their operations compared to similar groups in urban or metropolitan settings. During periods of economic downturn in single industry resource towns, voluntary groups often find themselves responding to increased demands. In contrast, rural and small town places that have multiple resource sectors may not experience singular boom-and-bust cycles, but rather regional waves as different sectors experience boom and bust at different times. In this context, multi-service agencies may be simultaneously responding to both the impacts of a boom-and-bust cycle across various resource sectors (Ryser et al. 2014).

Another key difference, and considerable challenge, is 'distance'. There is distance involved in connecting with governmental agencies or key policy makers that may determine the fate and the success of voluntary groups and / or their funding applications. Distance exacerbates the possibility that policy makers will not understand the context and circumstances of remote resource-dependent communities. The opportunity to connect with wider networks of organizations, or with government ministries and regionalized supports, is also negatively impacted by the friction of both social distances, which are created through increasingly lengthy bureaucratic processes and accountability procedures, and the physical distances within which these voluntary groups operate. Distance also impacts access to information and different types of innovations, ideas, options, and solutions. With limited external networks, voluntary groups can become isolated and introverted; thereby limiting their capacity for renewal (Wollebæk, 2009). Organizations in resource-dependent regions also know that a significant portion of their annual operating budget may be consumed by travel costs – costs either to deliver services in a wide-reaching geographic area, or costs associated with connecting with urban-based policy and program offices (Harris et al., 2004).

Trends

Having set out a foundation for understanding both civil society and the voluntary sector within the context of a staples-dependent economy, this section of the chapter explores other trends impacting voluntary-sector groups and organizations.

Contracts

In an effort to address local service needs, and in response to gaps left by profit-driven private enterprise, the state has turned in recent decades to the notion of funded partnerships with voluntary organizations. Recognizing that by not delivering services directly, the state has achieved cost savings, and also recognizing that there are costs for voluntary-sector groups to organize themselves and deliver social welfare services, the state has redirected some of its cost savings into new funding programs designed to support voluntary-sector groups in the delivery of such services. While details vary by jurisdiction, the general model is that the state identifies a service area need, puts together a funding program and calls for proposals from groups with not-for-profit status, then awards the contract and monitors its implementation. In addition to the general problems with government funding programs around their increasingly burdensome reporting procedures and their changes in focus as a result of political machinations, this approach creates some distinct issues and challenges for voluntary-sector organizations.

To compete for state service contracts, voluntary organizations often must 'up their game'. This includes having an increasingly professional and trained staff, a well-ordered organizational structure, comprehensive board management and training, senior leadership, and the associated facilities and infrastructure to support service delivery. These needs start to generate the funding treadmill where fiscal resources are needed to build the organization, with the organization needing to grow to keep winning state service provision contracts, which in turn means needing more contracts to pay for the organization's operating costs, and so the cycle goes. The transition to larger state contracts with larger and better-organized enterprising non-profits has also meant increasing competition between such groups. Very often, this competition countermands the hopes of state funders that voluntary organizations might partner as these organizations themselves get into battles for survival and funding contracts.

Scale and scope of operations

Voluntary and civil society organizations struggle with questions about the scale and scope of their operations at both a local and regional level. As the number of needs or contracts expanded, the scope of their operations typically became more complex as organizations were delivering programs at multiple scales. In addition to problems associated with 'mission creep', the cycle of pursuing new funding contracts means that voluntary and civil society organizations often needed to expand their topical and geographical mandates into multiple service areas in order to remain viable and to address emerging service gaps.

If successful with state funding competitions, the service area for resulting contracts often corresponds to the service areas of senior government agency offices – be that focused within a single community or spanning across an entire region. With an increase in referrals and demand for services from nearby communities, voluntary organizations may also have expanded their mandates to deliver services at a regional level. Changes towards a regional scale and mandate have also been shaped by new collaborative and regional service delivery models that are being advocated for by senior governments.

This difference in scale can affect levels of resources and structural supports available to voluntary-sector groups. It also links to our earlier discussions of the importance of bonding and bridging forms of social capital in order to strengthen the capacity of rural organizations. Building upon this, civil society organizations, including voluntary groups, have been pursuing opportunities for improved financial efficiencies and better economies of scale by engaging in shared services and infrastructure arrangements (Paagman et al., 2015; Walsh, 2008). The ability of an organization to significantly reduce costs can be difficult. This has prompted organizations to consider scaling up, in order to purchase bulk equipment and supplies. There are also potential savings for organizations by sharing equipment, vehicles, and staff such as IT, payroll, human resources, and reception supports (Lennie, 2010). While they often start on one small issue, organizations often find through the exercise of their activities that their networks link them to a wider array of organizations or supports.

New mobile labour landscapes and civil society

As noted earlier, resource towns that were once built to accommodate large local workforces are now immersed in much more fluid flows of labour and capital (Haslam McKenzie and Rowley, 2013). Following the global recession of 1982–1984, government and industrial restructuring focused on shifting away from building new single industry communities in rural regions (Storey, 2010). From an industry perspective, improvements in (and long-term cost reductions to) transportation and communication, the adoption of flexible production techniques, the adoption of extended shifts to support year-round and 24 hours per day operations, and access to a larger supply of qualified workers also helped to make mobile workforces more appealing (Rolfe and Kinnear, 2013; Tonts, 2010). While Staples theory and comprehensive planning approaches once conceptualized localized and traditional labour markets, restructuring of these industries, declining job benefits, trends towards short employment contracts, and limited access to nearby resoruces, has precipitated and reinforced mobile labour in competitive regional, national, and global labour markets (Ryser et al., 2016b). These new labour arrangements have profoundly impacted civil society in many ways.

Long-distance labour commuting (see Chapter 11) has generally diminished the volunteer base available in small communities. As residents engage in long-distance labour commuting, their engagement with voluntary organizations in their home community may become what Wollebæk and Selle (2004) describe as more random, less frequent, and non-committal, thereby impacting the ability of such voluntary groups to maintain their operations and as well as the local social cohesion and social capital needed to remain effective.

Aging-in-place

In addition to population decline and labour-mobility pressures, the remaining population is often aging. This trend is both attributed to the out-migration of younger families and the presence of an older workforce that is aging in place. As identified by Hanlon and Halseth (2005), such 'resource frontier aging' can limit community capacity to sustain ongoing initiatives. These issues exacerbate the long-standing challenge of leadership renewal for voluntary groups in rural and small town places (Skinner and Joseph, 2011). By definition, small places have small populations and this means that if leaders move on from their position, or burn out from the stress of service delivery, there are relatively few potential volunteers to draw from. There may also be limited participation by members. While older volunteers remain a valuable resource of voluntary-sector groups, some may leave the community over the winter months; leaving a gap or interruption in their volunteer commitment with implications for maintaining momentum for organizational initiatives.

Aging building / infrastructure assets

The abundance of aging infrastructure in non-metropolitan areas manifests important issues for renewal. As a result of industrial restructuring processes, small resource-dependent communities tend to have many non-profit organizations and groups that have limited resources. While these groups benefit from being able to lease or own affordable infrastructure, it has also often left these small groups dispersed across many tiny and inadequate spaces. There are missed opportunities to create synergies, renew infrastructure, and reduce operational costs through shared space.

Rapidly growing resource-based communities have also experienced unique infrastructure renewal challenges of their own. Despite similar concerns of aging infrastructure in these communities, pressures from renewed industry activity can made it very difficult for voluntary groups and organizations to acquire property or space to support service delivery or shared uses. Rapid growth associated with industries such as mining and LNG, however, can also produce dramatic and sudden increases in operational expenses for non-profit agencies as landlords seek to take advantage of industrial expansion via increased lease rates.

Responses

In this final section, we want to discuss how voluntary-sector organizations in remote, rural, and resource-dependent places have developed practical responses in order to remain viable and to continue to address the needs that they see on the ground in their communities. We focus on how they are diversifying their human and financial capital, as well as how they are using partnerships, smart or shared infrastructure, and scaled up approaches to support organizational renewal and ongoing operations.

Human capital

To start, voluntary organizations typically build their human capital through diverse recruiting strategies and joint recruitment campaigns (Walk et al., 2013). Some groups have offered compensation for items such as supplies, memberships for emergency roadside assistance, or fuel, which can be a significant cost in rural regions. Some communities have developed a family-friendly certification program that has been adopted by voluntary groups, as well as other public- and private-sector groups, to recruit new members, volunteers, and staff. The goal of such programs is to support flexible scheduling, access to child care, organizing family-friendly activities, and other things.

Small communities have also been looking at other volunteer rewards and incentives, ranging from passes to recreation centres to donations for community groups.

Organizations have become more flexible with the types of commitments that groups are looking for when they recruit new people. This has been especially important during times of significant industrial restructuring, where job losses may have prompted much of the local labour force to commute long distances for work (McDonald et al., 2012a). Emergency services, such as volunteer fire departments and search and rescue teams, often require volunteers to regularly attend training exercises. This is not always possible, with rotating shift schedules and out-of-town work. Voluntary organizations have been flexible in finding ways for these workers to continue their training and engagement.

Voluntary groups are also exploring appropriate opportunities for cross-training, joint training, mentoring, and succession planning in order to provide coverage for delivering programs and to strengthen risk management plans. Some senior government agencies, as well as non-profit organizations, have developed workshops to provide training and advice to voluntary groups. Training and professional development has also been used as incentives to recruit and retain volunteers (Walk et al., 2013). In some cases, voluntary-sector organizations in resource-dependent communities have worked to build the capacity of low-skilled clients by engaging them in organizational tasks, such as data entry or program operations. This not only equips their clients with the skills they need to better engage in the workforce, but it also provides the organization with more human resource capacity.

Having stable and adequate human resources can really impact the resilience of an organization. Diverse human resources, for example, provide stability for developing and maintaining funding and partnerships (Milbourne et al., 2003). Diversifying human resources and networks across different economic sectors in the community, and across different ages, has been important. Some voluntary organizations, for example, have used their networks to expand their human resources by subcontracting or sharing administrative, financial management, and fundraising staff (Poole et al., 2002). Voluntary groups are also completing human resource strategies to examine wage parity issues and appropriate opportunities for staff development. 'Blended' positions (i.e. combining tasks from multiple contracts in order to create full-time jobs) also help to improve recruitment and retention rates for staff in small resource-dependent places.

Financial capital

Voluntary organizations are also diversifying their funding sources by pursuing a broader range of funding and in-kind resources from local and senior levels of government, industry, business, trusts, and other sources. Local governments, for example, have provided financial support through amenity fees, grant-in-aids, permissive tax relief, and low-cost lease arrangements. They have also provided in-kind support for installing equipment, drafting infrastructure agreements, providing advice on design and construction, and brokering relationships with senior government and industry partners. Voluntary organizations have also worked with senior government to transfer properties at nominal costs in order to support multi-purpose initiatives. Some organizations have also been able to obtain capital grants from senior government agencies that enable them to replace lease arrangements with mortgages and strengthen the stability and resiliency of social infrastructure in small communities.

Industries continue to invest in small places through community donation programs, but it is often their in-kind support, through engineering or construction expertise, donated materials and labour, and donated equipment, that is most valued (Ryser et al., 2016a). Voluntary groups have further diversified their financial resources through investments, property management activities, and through alternative sources of financing. The challenges of the private-sector model approach in rural and small town places has, in particular, supported a new form of voluntary organization – the enterprising non-profit group. While various forms of social enterprises have long existed, and that the notion of community or social responsibility as an integral and intimate part of 'economic' activity has for millennia been part of Indigenous / Aboriginal approaches, there is more pressure for non-profits to grow their own revenue streams in order to sustain ongoing operations. The development of social enterprises in rural communities is still largely emerging, but there have been some important role models (Ryser and Halseth, 2012).

Partnerships

Partnerships are increasingly important to provide supports to vulnerable groups in resource-dependent regions that are complex and often beyond the capacity or mandate of any individual organization – something that is particularly important in smaller communities with fewer specialized services and resources. In remote, rural, and resource-dependent places, organizations are investing more time to strategically identify and understand the capacity (assets, skills, strengths, networks, potential synergies, threats, liabilities, and weaknesses) of potential partners. In many cases, sub-committees or small-scale and limited duration projects are used to test cooperative working relationships and build trust before pursuing larger joint or shared initiatives. Integrated service teams have also been used to explore possible efficiencies, synergies, and capacities to manage and deliver services.

Voluntary organizations have also been using partnerships to renew relationships with various stakeholders (Ryser and Halseth, 2013). This has been particularly important when these relationships involve Aboriginal and non-Aboriginal groups where the key to relationships and capacity is flexibility in terms of how they work together. Furthermore, voluntary-sector organizations have been working to collaborate and align messages with local government for when they engage with other industry and senior government stakeholders (Ryser et al., 2012b). In some cases, they are also working to find projects that will be of interest to industry in order to enhance their relevance in renewal activities (i.e. water conservation, recreation, health).

Shared infrastructure and service arrangements

Voluntary organizations in remote, rural, and resource-dependent places are paying particular attention to infrastructure needs and arrangements to maintain supports that might not otherwise exist. Small communities are increasingly confronted with the challenges of aging and inadequate infrastructure – much of which was developed during the immediate post-Second World War-era of resource frontier expansion. Many of these structures require new foundations, roofs, windows, insulation, and other energy-efficient considerations. Unfortunately, the closures of schools, former military buildings, businesses, and industries over time have left small communities with poorly maintained and underutilized infrastructure. The limitations of these structures are not only that they are aging, but they were never designed for uses envisioned today. At the same time, older and unrenovated infrastructure may not be accessible for young families with strollers, people with disabilities, or an aging population. Such challenges are not unique to rural organizations, but they can exacerbate pressures and undermine the effectiveness of an already limited capacity in underserviced small communities. Major renovations are needed to update rural facilities and equip them with the technology, equipment, and space to support broader community development initiatives. Investments in technological infrastructure, such as high-speed internet and audio-visual capabilities, have enhanced the capacity to support requests for joint training sessions.

In response to limited or aging infrastructure issues, local governments and the private sector have provided free or low cost access to meeting or operational space for voluntary organizations (Ryser and Halseth, 2013). Churches in small communities are also increasingly playing a role in supporting social housing investments and infrastructure for voluntary organizations. This has been important, of course, in communities struggling with temporary or permanent industry closures, but also in booming resource communities where commercial vacancies are low and commercial rental costs are rising. More organizations in smaller communities are addressing these infrastructure challenges by co-locating or developing multi-purpose facilities in order to develop synergies, collaborate, and enhance communication across service providers (Moseley et al., 2004). This can also provide a more efficient portal for residents to access information about needed supports. In resource-dependent regions, multi-purpose facilities have also benefited from in-kind contributions by residents and industry.

Scaling up with a regional approach

Local voices and organizations are scaling up and experimenting with new institutional structures and relationships in response to the withdrawal of senior government supports, as well as to search for more streamlined and innovative approaches to provide services that otherwise would not exist in these more remote resource-dependent communities (Smyth et al., 2004). A critical challenge remains, however, that many regional approaches and structures are underfunded, ill-equipped, and underdeveloped. Stronger regional systems will only evolve as a key mechanism for renewing resource-dependent regions, if they are adequately resourced and empowered to engage in long-term strategic planning.

While much of the focus in community development contexts has been on horizontal relationships, it is important for non-profit and voluntary initiatives to include vertical, or extra-regional, relationships with multi-level political connections in order to enhance access to resources and support across a broader range of stakeholders (Shucksmith, 2009). Underdeveloped synergies across community, regional, and senior government stakeholders, however, reflect a lack of financial and political capital to be effective and mobilize change. In Canada, challenges strengthening vertical relationships have been compounded by the uncertainty created by the federal government's transfer of responsibilities (such as in some housing and employment programs) to provincial authorities, as well as the provincial government's decision to reduce the number of contractors and privatize some programs. In many cases, government agencies responsible for distinct sectors have retained power and limited how civil society actors can respond to new opportunities and approaches or build upon a unique regional context and rural assets. Such policy approaches reinforce path-dependency in rural areas. While senior governments have called for more coordination, partnerships, and integrated supports, voluntary organizations been challenged to implement top-down operating protocols, accountability, and collaborative operating procedures – all of which have further limited the ability of staff to be flexible and responsive to the needs of rural residents.

Conclusion

As remote, rural, and small town resource-dependent places continue to manoeuvre the constantly changing pressures and opportunities of the global economy, voluntary organizations are becoming increasingly valuable assets because they are able to mobilize various forms of community capital in support of diverse community and economic development initiatives. In the competitive global economy and within a Neoliberal policy framework, civil society itself, in many ways, represents the cornerstone of local and global spaces, where advantageous positions are continuously being negotiated and renegotiated. They are providing opportunities for local and senior governments to mitigate expenses and balance their budgets. They are strengthening the competitive advantage of industries by helping to recruit and retain labour, while supporting the monitoring and legitimacy of their operations in global markets.

While voluntary organizations are increasingly important to sustaining communities, their new role and place is often misunderstood, poorly supported, or undervalued. Voluntary organizations are no longer dominated by informal governance and operational processes, but are increasingly professionalized, even business-oriented. This transformation requires investments commensurate to renew and retool the capacity of these organizations to pursue new service and infrastructure arrangements at the local and regional level. Policies and restructuring pressures that are driving new expectations for voluntary-sector groups, however, are challenging the transformative capacity of rural organizations. While such challenges are not unique to rural organizations, they can exacerbate pressures and undermine the effectiveness of an already limited capacity in underserviced small communities. With limited human, financial, infrastructure, and political capital, voluntary-sector organizations must ensure that these limited capacities are not wasted, but are purposefully deployed in the most relevant and effective way. As voluntary organizations continue to assert a key role in renewing remote, rural, and resource-dependent communities, there is a void of appropriate governance, infrastructure, human resource, funding, and policy tools that requires attention if they are to more fully deliver on their potential.

11 Individuals

Introduction

This chapter shifts the scale of our discussions regarding the changing political economy of resource-dependent rural and small town places. In turning our focus to 'individuals', we overtly recognize that people are not only experiencing the impacts of changing economies and changing societies very intimately, but also that individuals are at the very centre of a political economy analysis. Without the support of individuals, the economic and political supports that define political-economic structures at any given point in time will not be sustainable. As a result, a contest of wills and coercion begins, with individuals expressing their hopes and desires and capital (working through economic as well as political structures) seeking to mould and shape those hopes and desires. In turning our attention to individuals, we also provide a lens through which the implications of broader changes and trends can be evaluated.

The chapter is organized into three sections. The first provides some background to the study of individuals within the social sciences. In addition to historical approaches to such studies, the background section also describes some of the theoretical positions and debates important in shaping how studies of people within dominant political-economic systems have been, and can be, carried out. The second section turns our attention to the individual in the economy and in society. The section starts with a review of how changes in the economic opportunities of resource-dependent rural and small town places impacts individuals. It then moves into a review of the impacts and implications of wider economic and social changes on households. Included is consideration of changing gender roles and expectations. The third section draws out key trends impacting individuals within rural and small town places. Among these trends are demographic aging and the increasing use of mobile workforces within resource industries.

Background

Social geography's research interest in the individual as a focal point for studying the processes and impacts of change has a long pedigree. As noted in the introductory chapters, some threads link back to human ecology as practised at the University of Chicago. Beginning in the 1920s and 1930s, and under the direction of Robert Park, the 'Chicago School' emphasized ethnographic fieldwork and deep immersion in daily life as they sought to describe and understand how people were engaging with the new realities of very large urban places.

Another set of threads link back to the theoretical ferment of the 1960s and 1970s in the social sciences. For example, the feminist critique of mainstream social science methodologies sought to include (among many things) both the researcher and the research subject, not as participants, but as gendered human beings engaged together in both the research project and the life that surrounds the research project.

Our interest with individuals also links to those taken up by the humanist critique of that same period. Focusing on re-introducing people in the quantitative modelling or structural explanations that were vying for supremacy through the 1960s and 1970s, the humanist critique sought to shine light on complexity and on context.

Within rural geography, Paniagua (2016) identifies that much of the work has been focused at the 'community' level. This has been partly a reflection of the tremendous changes that have been impacting rural places and communities. But what about the individuals within these places? As Paniagua (2016, p. 513) points out, "Different members of communities perceive and experience the community in different ways". Just as Massey had suggested that places respond to change in their own unique ways as a result of the uniqueness of each place, so too will individuals.

Our concerns are with the ways in which individuals connect with, and are impacted by, the various institutional structures introduced in this part of the book and which play a role in maintaining and regulating the political economy of capitalism. That the uneven geographies of the capitalist economy extend to the individual person is well described in many different research settings. For the individual, "the unequal distribution of opportunities" (Ley, 1983, p. 278) can limit access to employment, housing, services, and more generally to quality of life.

The individual in the economy and in society

This second section of the chapter emphasizes individuals within the context of a changing economy and changing society. We know from previous chapters just how dramatic the economic transitions impacting rural and small town resource-dependent places have been. How people have been affected by this means going beyond questions of jobs and job losses. As a result, we also explore how the changing nature of work has affected gender roles and relations, as well as individual identity.

The second part of this section looks at some of the trends evident in the changing nature and structure of households. The historical and stereotypical resource town household with the male partner at work in a mill or mine, and the female partner at home raising the children, was only ever that – a stereotype – but it has also changed in complex ways. Economic restructuring has meant that while employment opportunities of some types have closed off, other opportunities have opened up. This has created opportunities for a more symmetrical construction of households and families.

Changing economies

The structure and organization of this book reflects an understanding that significant economic change and restructuring has been a driving force in shaping and reshaping rural and small town resource-dependent places for more than 40 years. These forces will continue to reshape these places for many years yet to come. In this part of the chapter, we explore some of the impacts and implications of economic change and restructuring for people in resource-dependent communities.

Job losses

To start, we know that the imperative to produce a profit or return on investment lies at the heart of the capitalist economic system. In the post-war period, the opportunity to apply Fordist industrial organization models to resource development in the rural and remote regions of developed economies created large numbers of jobs that absorbed surplus labour and created the positively recursive production-consumption relationship that supported the post-war 'long-boom' economic cycle. With the challenges of globalization, resource industries followed their manufacturing cousins in programs of economic restructuring. Faced with long-term downward pressure on commodity prices, and accelerating competition from low-cost resource-producing regions, these cycles of restructuring also accelerated with the result that there were significant employment reductions in various resource sectors.

These job losses had significant impacts on individuals and families. In sectors that had long enjoyed a relative level of job security, workers could suddenly find themselves unemployed. In this context, resource workers not only experience a loss of income, but may also encounter prolonged periods of unemployment, loss of benefits, lower living standards, and impacts on their retirement plans. A first consequence of such a change is stress and the associated negative cycles of physical and substance / alcohol abuse. In resource-dependent places with massive resource-sector layoffs, service providers often struggled to cope with an upswing in requests for assistance and support.

Another outcome witnessed around resource-industry layoffs and closures is a measure of 'denial' by workers. Often these resource regions will have experienced past up- and downswings in activity due to global market prices and demands. Having recovered numbers of times in the past, it can create a false hope among unemployed workers that they just need to 'hold on' until the markets change and then the mill or mine will re-open. Unfortunately, the permanent closure of jobs can leave these individuals spending the last of their money just to keep holding on which leaves them without the resources to move or take training for new jobs.

The consequences of being laid off or without employment are difficult for many workers, but the history of post-war resource sectors jobs in developing economies has created some specific challenges. One of the first is that resource-sector jobs were high-wage jobs and individual and household spending often aligned with that wage scale. As noted in earlier chapters, resource towns are characterized by a significant dichotomy between high-paying resource-sector jobs and the rest of local employment. Unemployed resource-sector workers find it difficult to obtain employment locally due both to limited opportunities and to their often-unrealistic wage expectations. With fewer employment offices in rural and small town places, scarce information about wider labour market trends can also affect employment opportunities and decisions (Levernier et al., 2000).

Another consequence is that resource-industry jobs were historically geared to heavy labouring work and many got jobs with limited education or formal training. As jobs became more complex, whatever training was needed was learned on the job. To become unemployed in late middle age, and to have relatively limited formal education, leaves many of these workers in a difficult spot should they try to obtain work in other fields. Skills transfer between resource sectors is even very difficult. Some researchers argue that an overspecialization of skills is also impacting the ability of residents to adapt to changing economic conditions. For example, Freudenburg and Gramling (1994, p. 11) argue that "when the offshore oil industry is in decline, an argon welder's skills may provide little in the way of the kind of human capital needed to get another job".

As displaced workers seek to re-enter the workforce, they are placed in a weaker bargaining position – with older, less educated workers facing greater difficulties. When workers in households suffer economically from unemployment, coping strategies vary, and these will be followed up in the next section of this chapter.

Identity

To this point, the impacts discussed have been economic and employment related. As we know, employment also contributes to the identity and sense of self-worth of workers. The identity of resource workers is strongly related to their workplace or sector (Major and Winters, 2013), to their community, to their role in their family or social structure, and also increasingly their mobility (Dufty-Jones and Wray, 2013). Historically, companies supported many events for workers and their families. These events were designed to strengthen cohesion and a sense of belonging that would in turn strengthen workplace synergies and a commitment to the company and community (Halseth and Sullivan, 2002).

To lose a job, especially one that commanded a high wage in the dominant local economic sector, has an impact on the individual and their sense of identity. After decades of labour market volatility and becoming unemployed, older workers can lose trust and can question their identity as resource sector workers (Lassus et al., 2015). Workers can feel betrayed by industry after working long hours and years to improve productivity (McDonald et al., 2012b). Jobs losses, prolonged unemployment, and prolonged absences through long-distance labour commuting can lead to poor nutrition, lack of confidence, isolation, loneliness, depression, anxiety, substance abuse, violence, strained relationships, and even suicides (Brand, 2015; Lassus et al., 2015). Through periods, workers can feel stigmatized; even branded as someone who lost their job because they were somehow less productive. Their loss of identity can be further impacted by long-distance labour commuting – where workers' sense of community identity is diminished by prolonged absences that limit their involvement in organizations and activities. Research in Australia suggests that international mobile workers have struggled to regain employment in their home countries as their pursuit of international mobile work meant they were no longer 'loyal employees' (McDonald et al., 2012b).

A further element of personal identity that is challenged with unemployment comes from the loss of routine and structure in one's life. Workers lose their daily structure, their ability to demonstrate skills, and a deterioration and loss of their social network assets (Lassus et al., 2015). The discipline of the clock, of work shifts, of weekends, and of annual vacations is lost and each also removes the anchors to both activity and the identity we create for ourselves around our work ethic and our position in the community. As Wetzstein (2011, p. 2) argues, "Rapid economic and demographic changes associated with large-scale resource development was understood to lead inevitably to social and psychological dislocation and a breakdown of established community social structures". Many friendships and social networks that had been based upon local activities and events are no longer possible to maintain, leaving workers and their families more isolated (McDonald et al., 2012b). Job losses have altered the social structure for resource workers, as they are no longer the primary income earners for their families. As family relationships have become strained, job losses of resource workers can also result in higher rates of separation and divorce, higher suspension and school drop-out rates of children, lower levels of education for children, residential mobility, and disruption of children's school and social networks.

Finally, one of the most challenging personal issues associated with unemployment is the loss of camaraderie with one's co-workers. For many, this cohort of workers acts as a reference point through which the issues in our lives and in our interpretation of the world around us are discussed. The norms, cultures, and behaviours of the work place play a role in helping to define us. The workplace coffee-break, for example, also forms a social venue for making sense of issues and the asking of opinions. This loss is also felt by retirees and the coping mechanisms are dependent upon how such workplace social arrangements can or cannot be replicated outside of the working environment.

Job loss and unemployment impacts differ for each individual depending upon their coping abilities and their resilience or preparedness. It is important to note that impacts also differ by age. As just noted above, older workers can have a very difficult time seeking out new jobs when their skills or education may not match those needed in other sectors. Younger workers are vulnerable to especially sharp economic consequences. Housing mortgages, vehicle loan payments, and consumer debt may have all been manageable when one is working full-time at a high-paying resource-sector job. But these things can quickly force individuals into bankruptcy or result in the loss of those assets when that high-wage employment is suddenly terminated. As many young workers are also in their family formative years, the impacts of a loss of income affect everyone in the household.

Changing nature of work

In addition to the issue of absolute job losses, there are also numbers of changes in the nature and organization of work that are impacting individuals. As described earlier, one of the processes embodied within restructuring has been the spinning off of full-time workers into casual or contract labour positions. This casualization of the workforce changes the dynamic of work and of job security. When work is reduced from full-time to part-time or casual, many of the impacts described above for unemployment (wage loss, impacts on identity and social networks, etc.) are replicated. When work is contracted out, it may come with new opportunities to be entrepreneurial, to set one's own hours, and to get paid for what the work is worth, but it also comes with risks and vulnerabilities.

A further consequence of the substitution of capital for labour and ongoing processes of restructuring is that the nature of the work that remains in resource industries has also been changed. In mills and mines, for example, large shifts and crews doing heavy labouring work with the natural resource itself is often replaced today by a small number of workers operating technology from remote locations or isolated control booths. Instead of placing hands on ore or wood, the job is about monitoring computer screens and making process adjustments when sensors highlight potential problems. High levels of training, an ability to do more tasks in the production process, and the context of working both individually but also as part of a speciality production team are all consequences for the drive for labour-force flexibility.

For workers already in the resource sector, there are usually opportunities for progressive skills training as new processes, new technologies, new equipment, and new jobs are introduced. For young people, these changes mean that the opportunity to get a high-paying and secure local job in the resource sector after dropping out of school (such as may have been available in the 1960s or 1970s) no longer exist today. Contemporary resource-sector work is increasingly for specialists with high levels of education, training, or particular skills / trades certificates. During a recent economic boom in northern British Columbia, unskilled workers flocked to the area with hopes of getting work, but the human resources staff was forced to tell them all that they 'cannot even walk onto the job site' without at least three or four certificates in first-aid training and workplace health and safety.

In addition to the changing need for education and training, and the commensurate capacity for ongoing learning, the change in the nature of resource-sector work from heavy labouring to more technical and monitoring work has had at least two impacts on new entrants to these sectors. The first is that resource-sector work had historically been one place where new immigrants could be successful in finding good employment. The limited numbers of jobs in many resource sectors, except where there might be periodic booms, and where specialized skills sets are required, now limit some of these opportunities. On the other hand, the changing nature of work has assisted with the breaking-down of gender barriers in many resource sectors. The opportunity for women to participate is something of a break from the past and both changing social norms and values, together with the changing nature of work and the workplace, play a role. Despite such changing opportunities, this does not suggest that women's participation in resource-sector jobs is not without difficulties. To start, sexist attitudes can still persist in any employment setting and these can create unwelcome workplaces. Perhaps more limiting, however, is simply that so many resource sectors have fewer jobs per unit volume of production now than in the past. There are fewer job openings, and layoffs of the most recently hired is potentially likely whenever there is an economic downturn or when new technology is introduced to make production more efficient.

Children and young people

To this point, the discussion of restructuring impacts on individuals has focused on adults. However, children and young people are also impacted by these changes and too often those impacts are not addressed by transition support services. To set a context for understanding the impacts of restructuring on children and young people, we need to imagine them living in their homes with their parent or parents. Stresses on the parent are likely to be felt by all members in the family. If stress leads to conflict and various forms of abuse, the impacts felt by all members of the family become even more significant.

As if experiencing these stresses is not challenging enough, children and young people do not have the coping mechanisms with which to deal with these impacts. For some groups of children in particular, such as teenage boys, difficulty in sharing their feelings or talking about their experiences makes it hard to get the assistance supports they might need.

In addition to the general stresses experienced within the household, children are also impacted when a parent has to leave the home for some form of long-distance labour commuting. This may not only upset household routines, but it can also change the relationship between the child and the commuting worker. In one resource-dependent town where long-distance labour commuting became a common activity following the closure of the local industry, one school principal tells a powerful story:

> A teacher went out into the hallway and saw a young child sitting on the floor crying quietly to himself. The teacher went over to see what was the matter. The young student said that he had just had a wonderful weekend with his father, they had done all sorts of fun things. But his father had left this morning to fly out of town for work and he wouldn't see him again for ten days. This made him very sad and was the reason he was crying. The teacher tried to console the young student. She said she knew what the student meant. Her husband was back in town last week for four days and they had a very nice time. They had dinners together and went for walks, but then he had to go back to work and she won't see him again for another two weeks. A short time later, the school principal came into the hallway and saw both the young student and the teacher sitting on the floor both crying quietly to themselves.

Another significant impact for children occurs with out-migration. When local jobs are lost, many households need to move to take up new job opportunities. This can mean a loss of childhood friends as other families leave the community – or especially the loss of all of their friends when the children's family themselves has to leave the community. Depending upon the age of the children, such moves may be more or less stressful.

For young people in the community, the changes brought by economic restructuring and industrial downsizing have a big impact on what was once one of the most lucrative parts of growing up in these towns – the summer job. Historically, summer jobs were plentiful as young people filled in for the workers taking their vacation time. Hiring practices often favoured local youth, especially those local youth whose parent / parents worked for the company. The wages also tended to be very good. With a reduction in overall employment and transitions in production cycles, as well as the more stringent safety and training requirements in place, many of these summer replacement jobs have been severely curtailed.

Changing households

Together with economic change and impacts focused upon work and the workplace, individuals in rural and small town resource-dependent places have also been impacted by our changing society. With the household as a focal point, this section of the chapter looks at the historical construction of households and household roles, and then updates that historical portrait to trace two eras of household adjustments – the immediate contestation of dramatic economic restructuring, and the more recent discourse on symmetrical households in resource-dependent regions.

The household has long been a fundamental unit of analysis in social geography and the other social sciences, and studies of the relationship between gender and work have tracked wide swings in the roles and contributions of family members to the household economy. In the pre-industrial period, Newby (1987) illustrates how all members contributed in a variety of ways to the survival of the household in both town and country settings. During the early period of industrialization, studies from the United States and Britain highlighted how women were the first to take up factory jobs and thus contribute cash wage earnings to the household economy (Dublin, 1979). After the Second World War, the redefinition of gender roles at home and at work created not only the stereotypical 'pink collar ghettos' of 1960s and 1970s suburbia (Jackson, 1985; England, 1993) but also set the foundation for what would become acknowledged as hegemonic male urban planning constructions such as the 'journey-to-work' (Hayden, 1995). The theme linking these very separate research directions and subject periods is that they collectively highlight the restructuring of household roles in response to economic change.

Traditional understandings of the social geography of single-industry and resource-dependent hinterland towns have portrayed them as being highly structured by gender (Egan and Klausen, 1998). In the periods both before and immediately after the Second World War, the workplace was conceptualized as male space while the home was similarly defined as female space. Such actions fit with more general notions of the period where men were viewed as economic providers while women were seen as the primary care givers and nurturers (Gibson-Graham, 1996). This gendered separation of productive and reproductive spaces defined life and lifestyles, as well as much of the social and economic research on these towns.

The post-war model of resource town households was of a male breadwinner, female homemaker, and several young children. A product of both the social and economic structures of the times, these post-war resource town settings magnified stereotypical roles. As described by Halseth (1999), these households came for the work opportunities that the post-war expanding resource-development sector was creating.

Forestry, mining, and farming jobs involving the operation of equipment and machinery were described at the time as 'men's work' and were seen as heavy, dirty, dangerous, as well as requiring physical strength and long hours of labour in harsh environments (Little, 2002). Social and technological changes have challenged this description, however, and are breaking down barriers to women's labour force participation (McDowell, 2001). However, writing with respect to a set of BC forestry towns, Reed (1999, p. 4) highlights how "women's contributions – both to the paid work of the forestry industry and to the unpaid work of maintaining forestry communities – are many, yet they remain unacknowledged".

Coupled with the post-war baby boom, most rural and small town resource-dependent places became the settings for households with large families. This family formation period created many new household tasks. Support for women who took paid employment was limited. While

> the company began a new policy in the late 1970s of hiring women for the mines in non-traditional jobs ... the number of women employed (in the mines) in 1980 remains scarce enough that they were still regarded as exceptions.
>
> (St-Martin, 1981, pp. 144–145)

The nature and organization of the work, together with the very powerful social messaging about appropriate gender roles in the post-war world, supported this historical model of gender roles.

A gendered conceptualization of resource town landscapes had emphasized women's place as caretaker of the home. This conceptualization builds upon older constructions of gender roles but has been somewhat enduring for resource towns. Female domestic labour has not only played an important role in maintaining a reasonable standard of living for the household, but it has also been a foundation for the reproduction of the labour force (Lahiri-Dutt, 2012). According to Luxton (1980), such reproductive roles have included looking after members of the family, child bearing and child rearing, housework, money management, and shopping. Preston et al. (2000) found that regardless of paid employment, women are often still responsible for grocery shopping, food and meal preparation, laundry, and routine household tasks.

In addition to these household reproductive roles, women had also extended their domestic roles into the community (Little, 2002). Carrington and Pereira (2011) noted that women are more involved in community-based organizations and activities than men. For example, women have acted to improve the quality of life of their community by lobbying for schools, sidewalks, additional stores, community centres, and recreational facilities (Halseth and Sullivan, 2002). This is different from men's involvement, which tends to favour positions of power, such as the local government council (Little, 2002).

In resource towns, women's work historically played an important supporting role, with their lives often structured around their husband's employment (Sharma, 2010). This was particularly demanding under a shift-work regime, where women provide child care throughout the day, as well as attend to their family's needs, such as making meals, around the clock (O'Shaughnessy and Krogman, 2011). Shift work also means that women must keep the children quiet on days when their husband / partner needs to sleep (Preston et al., 2000). Through each of these examples, gender norms and expectations develop from, and are set within, the historical social relations of resource towns (Parr, 1990).

Life in resource towns also placed certain constraints on women's lives. With respect to household reproductive tasks, limited retail shopping choices and competition, together with high prices, affected a woman's ability to manage the family budget, particularly during times of economic crisis (Gill, 1990). Limited services may also have forced women to travel to larger centres, but this can be restrictive as many resource towns often lacked bus / transit service. Even where transportation is available, it could be logistically difficult and expensive for women to hire a babysitter for the time until they return.

Historically, there were few job opportunities for women in resource towns. Women's employment opportunities have generally centred on jobs in teaching, health care, retail, office, banking, and hospitality sectors. A 1976 advertisement for one experienced secretary in Flin Flon, Manitoba, received 124 applications from women (Luxton, 1980). Not only was there a limited choice of jobs, but these jobs were usually lower-paying and more likely to be part-time and without benefits (Hinde, 1997). Marchak (1973) noted that women had low incomes regardless of job control, education, or union status, and that their hopes for promotion are small compared to those of men.

The pay for such jobs has traditionally been poor and is relatively much lower than resource-sector jobs. In Flin Flon, for example, the company refused to hire women for underground work, and refused to hire women with children (Luxton, 1980). Reed (2003a) tells a similar story with respect to Port Alice, a forestry town located on northern Vancouver Island. Multiple barriers to women's employment on the shop floor of the local pulp and paper plant both restricted opportunity and reinforced patterns of hegemony. Only through a sustained set of legal procedures did women finally gain access to shop floor work. Despite optimistic reports of non-traditional job opportunities for women in Tumbler Ridge, BC, an early review of the Union of Operating Engineers membership shows only 84 women out of 1,150 members in the bargaining unit, a mere 7% (Peacock, 1985).

The notion of a male-dominated economic space is, however, an incomplete and inaccurate portrayal of resource town employment. While the historic pattern of differential access to paid employment continues, and the sense of masculinity associated with resource-industry work still finds local currency, the picture is more complicated. Reed (2003a, p. 382) suggests that:

> Very gradually, the numbers of women in professional forestry began to increase across the province. In 1995, (the first year in which data were segregated by gender) 10 percent of registered professional forestry graduates were women; by 2000, this proportion had increased to 14 percent.

Over time, of course, this stereotypical post-war model of the household started to break down. Of course, cracks in the model had always existed. There were lone-parent households, households where the female head worked, and changes in roles and responsibilities as households and families broke up or were recombined. The stereotypical model also started to break down as the economic challenges facing resource-dependent places accelerated after 1980. Job losses in the resource sector in particular challenged households to respond. As noted in a wide range of research, the responses of households to these challenges were multi-level. To start, some relocated for their work opportunities, sometimes within the same sector and sometimes changing economic sectors in order to secure work.

The historical and stereotypical household model has also been breaking down, due to some internal contradictions. Principal among these contradictions was the matching of education and skills with employment and income. In the immediate post-war period, the rapid growth of labouring jobs in various resource sectors were such that many young men could leave school early and secure good-paying resource-sector jobs. At the same time, young women would stay in school, often augmenting with post-secondary education, yet their job opportunities within resource towns was limited to service- and clerical-sector employment.

Job losses are one of the key consequences of industrial restructuring in Canada's resource-dependent towns. As individuals confront resource-industry restructuring at a very personal level, it is through the household that outcomes and reactions are experienced. Against a backcloth of employment losses, the need for survival of the household economy has meant a blurring of older notions and relations as gender at work and gender at home are mediated through both place and the needs and capacities of the household. Resource town households, therefore, play a central role in a community-wide shifting of the structure and understanding of gender roles and relations.

While locally constructed understandings of gender-segregated landscapes persist in resource towns, they fit increasingly less well with household realities that include dual incomes and shared parenting. Into this already complicated understanding is now added the upheaval inherent in the restructuring of resource industries. The result is transforming the social geography of both work and home.

Beyond traditional roles, Ali (1986) added women's more active participation in political protest as labour restructuring processes played out through layoffs, strikes, and lock-outs. This included picket line duty, involving both active as well as support roles – with the result that there were many arrests for breach of the peace. A second avenue of political protest included the occupation of government or corporate offices, as well as public events to generate publicity and media coverage.

Similarly, Gibson-Graham (1996, p. 208) refers to women as the backbone in strikes that "keeps their husbands' commitment from wavering". This is because women have an interest in their husbands' struggles at work, since his wage has implications for domestic work. In fact, women play an important role during strikes by walking picket lines, forming strike committees, support groups, babysitting cooperatives, car pools, potluck suppers, and entertainment (Hinde, 1997).

Key trends

Individuals in resource-dependent places and industries are impacted by the long-term trends that have been impacting both the places and the industries over time. While previous chapters in this book have described the impacts of job flexibility and economic restructuring on the form, nature, and number of resource-sector jobs, there are other more general trends that also impact individuals. Included among these are demographic aging and the rise of mobile work.

Demographic aging

The transformations that result from demographic aging have some very important impacts for individuals in rural and small town resource-dependent places. In this section, we review the historical trajectory of population demographics in such places, the impacts of economic restructuring on those trajectories, and highlight issues that will impact individuals – both young and old – in the coming years.

Single-industry and resource-dependent places in the hinterlands of developed economies have a rather distinct population story. While differences are seen across national contexts, the post-Second World War industrial resource model, coupled with the baby-boom phenomenon, has had a lasting demographic imprint. The post-war expansion of resource-industry development into more rural and remote areas, supported by new transportation infrastructure and industrial investments, all worked to create massive numbers of new job opportunities. Workers took up these resource-industry job opportunities, and buoyed by good-paying jobs and what was to become known as the long-boom economic cycle, they put down roots and raised their families.

Resource town development was geared to this young demographic. Civic infrastructure was designed and built for large numbers of young families who had large numbers of children. Parks, playgrounds, and schools dominated public-sector service investments. Such towns typically had few older residents, with the bulk of the population in the younger working-age cohorts. The nature of work in these towns structured the nature of economic and social opportunity. Social power developed within the highly gendered character of community landscapes, while economic power was focused on the resource industry. The price of everything from food to housing was priced at the wage levels of resource-industry jobs. This meant cost of living challenges for those employed in other, lower wage, sectors such as government, teaching, etc.

As the post-Second World War industrial resource model settled in, a curious thing happened. After experiencing a couple of decades of investment and expansion, the investment and expansion slowed – in many regions, it virtually stopped. With large resource industries in place, and with increasing efforts to make them more efficient and profitable, the massive job opportunities soon gave way to stable and secure employment for those already working. It also meant, however, very limited new job creation.

The lack of new job growth in the resource hinterland had a number of implications. As described by Hanlon and Halseth (2005), they all revolve around the process of 'resource frontier aging'. The existing workforce begins to age-in-place. Union protections ensured that as job losses arose due to the substitution of capital for labour, it was the more recently hired younger workers who were let go. At the same time as this workforce aging was going on, the lack of new local job opportunities meant an out-migration of young people. Sometimes they left to go to post-secondary education opportunities, sometimes for jobs, but the lack of local opportunity meant that they left. Recent studies in Canada (Halseth et al., 2017a) and Finland (Kotilainen et al., 2017) each highlight the hollowing-out of the younger age groups in the mature resource-industry towns of rural and remote hinterland regions.

These historical trends of in-migration followed by resource frontier aging create a unique set of contemporary challenges for individuals and resource-dependent small town places. In most such places, there are today fewer resource-sector jobs than in the past. While there are some 'booming' regions – such as those which supported the 2010–2015 explorations for natural gas reserves in a number of countries – most resource-sector jobs are occupied by older workers nearing retirement age. New jobs created in these places have generally not been sufficient in either number or pay scales to replace the lost resource-sector jobs. This context of fewer but older workers means that these places will be faced with significant recruitment issues for the first time in decades. Are the jobs and the communities – which were designed for workers and family expectations of the 1960s and 1970s – attractive to the current generation of younger workers?

Markey et al. (2012, p. 291) summarized some of the expectations in the literature regarding the next generation workforce. They highlighted that

> there has been a fundamental transition in what coming generations expect from the work and community life they will be going into. No longer bound by a primary concern over wages and jobs, young people are searching for quality of life, clean environments, and good places where they can raise children and achieve a balance between their work and life spheres. They want to live in places where they can make a dif- ference, every day, on a human and community level, while at the same time remaining 'plugged into' the global communications world.

For older workers, there are considerable issues as they step into their retirement years. First among these is the question of whether to stay in their resource town after retirement or to relocate. In favour of staying are the strong social and friendship networks they are likely to have built up over their working years. These networks are a critical part of healthy aging and provide important supports as people get older and need more assistance with the tasks of daily living. Men, however, often enter their retirement years with fewer networks of support compared to women; the differences often linked to the end of a marriage through separation, divorce, or death, but also through their greater propensity to social isolation, as many fail to develop supportive networks outside of work or their spousal relationships (Ryser and Halseth, 2011b).

Challenging the desire to stay are issues such as whether the community has the requisite services to support healthy aging. Such services include health care, retail, recreation, and the like. As these places were generally built for young families, they may or may not have the appropriate mix of services for an older population. In addition, the retrenchment of public-sector services described in Chapter 10 has left many places challenged in the area of health care services especially.

A second consideration is housing. The typical mix of resource town housing has historically been geared towards younger households. As a result, stairs are common, as are needs for considerable routine maintenance and upkeep of both the interior and exterior of the dwelling. In apartments, the typical three-storey buildings found in these communities often do not have elevators. Retrofitting is possible, but it can be expensive, depending on the quality and the design of the buildings. Such towns also typically have a limited supply of seniors' housing – especially given that historically there were very few older residents in the community. Selling a local house and relocating to a retirement community or to a city is an option, but it comes without the social networks identified above. Housing costs can also remain high in some resource regions. This places senior women living alone in a precarious position as they may have limited financial resources due to many years of lower wages, lower participation in the labour force, or possibly the loss of earnings or pension benefits from their former spouse (Ryser and Halseth, 2011a).

A third consideration is the design of the town itself. Is it organized in an age-friendly manner? Are there sidewalks and are they wheelchair-accessible? For that matter, are the public buildings, the shopping areas, and the recreation facilities designed for older people and for those with mobility constraints? Depending on the level of investment by the state and by local government in modifying the original designs, resource-dependent rural and small town places may or may not be amenable to aging-in-place.

Mobile labour

Spectrum

Mobile work is an increasingly discussed phenomenon, but it is also an increasingly complex phenomenon. Mobility has at least two aspects that we need to introduce. The first of these is the mobility from a "home" location to a "work" location (more on this below). The second involves those cases where the workplace itself is mobile. Included in the second part are jobs in airplanes, ships, trains, and the like. Also included are those jobs in distribution industries (delivery, courier, taxi, etc.) where the worker is in constant geographic motion during the day. Between these two aspects, of course, there is a gradation of both mobile work and mobile workers.

Another aspect of mobile work involves the geographic and temporal scale of the commute to work. At one end of the spectrum there is work at home.

This may be domestic labour or piece-work of some kind, or more commonly today it may involve some form of telecommuting or consulting via a home-based business.

Progressing along the commuting spectrum there is the daily commute model. Typically this entails leaving one's home before the work shift starts and returning to that home after the work day is over. This is part of the classic post-war commuting model and journey-to-work experience that supported the dramatic expansion of suburban living. Expanding the distance and duration of the commute then starts to get into a discussion of what is more commonly recognized as long-distance labour commuting (LDLC).

As part of a wider dialogue on the geographies of mobility (Kwan and Schwanen, 2016), our interest here is specifically with the topic of long-distance labour commuting within the natural-resources sector. From a worker perspective, Mertins-Kirkwood (2014, p. 1) defined labour mobility as "the free movement of workers between provinces, regions, or countries. ... High labour mobility is economically efficient, because workers can move to where they are most needed".

Markey et al. (2015) provide a description of LDLC in the Canadian context. While not a new phenomenon, contemporary LDLC has its origins in the offshore oil industry after the Second World War, and it has since become a common feature of the labour landscapes in Australia, North America, Russia, Norway, the Middle East, and many OECD countries. The business case for LDLC has been advanced by a number of political and economic restructuring forces (Storey, 2009). These include the cost of running work camps versus the extraordinary costs associated with new town development, including costs associated with increased environmental impact regulation. Restructuring within the energy and mineral sectors in particular has placed greater attention on productivity, reduced production costs, and the rationalization of unproductive operations (Tonts, 2010). Furthermore, while many workers and families enjoy the lifestyle offered by the isolated remote, rural, or small town resource-dependent places, with increased expectations for access to diverse employment, educational opportunities, and metropolitan lifestyles, LDLC provides companies with access to a larger pool of workers. Finally, companies may experience less difficulty in attracting and retaining workers using LDLC (Rolfe and Kinnear, 2013). Labour supply issues are exacerbated by the aging of the current workforce and limited success with attracting new entrants to replace retirees. In looking for new ways to attract and retain workers, in addition to salary inducements, companies are turning to provision of better quality accommodation and meeting demands for more flexible work arrangements which LDLC can offer.

Increases in LDLC have brought the issue of labour mobility to the attention of researchers and policy makers. Whether labour demand spikes due to natural resource 'rushes' around global development opportunities such as recently seen with LNG, or longer-term transformations in the relationship between increasingly mobile capital and its need for increasingly mobile labour, LDLC is being investigated and addressed in a number of contexts. For example, the intense pressures of LNG and mining development in Australia have prompted numbers of government inquiries and new research projects (House of Representatives, 2013; Nicholas and Welters, 2016). In Canada, regional disparities around natural resources (slumping fisheries in the east and booming oil / gas in the west) led to tremendous labour movement. Agreements beginning in 2009 increased labour mobility within the country as internal barriers to the movement of skilled or certified workers in a variety of occupations were systematically removed (Forum of Labour Market Ministers, 2015).

As noted, LDLC has gained increased industry practice since the 1980s. Storey (2010) describes the use of LDLC in a number of contexts and highlights the adjustment different industries have made to fit roster rotations to the needs of both the resource-industrial process and the workers themselves. Roster rotations may be relatively short, such as two to three days at the work site followed by two to three days off work. Roster rotations may also be quite long, involving 28–32 days at the work site and some equivalent time off work. More common cycles might be in between, such as two weeks at the work site, followed by one week off. Markey et al. (2015, p. 134) describe the complexity and diversity of work rosters across resource sectors:

> While LDLC operations share some basic elements, specific differences exist in the roster arrangements adopted. For example, offshore oil and gas projects are typically three weeks on/three weeks off (21/21) for those on rigs and platforms, 28/28 for those on supply vessels. Mine rosters show more variety with 7/7 standard for most workers at Saskatchewan uranium operations and 14/7, 14/14 and 21/7 commonly used at mines elsewhere. The construction sector has the greatest range of rosters influenced, in part, by the home location of workers. For example, Newfoundlanders working in Alberta on oil sands projects typically work 20/8, 21/7 or 42/14 rosters.

The cycle of time in and out of the work site is impacted by numbers of factors. Principal among these is the remoteness of the work site, because the travel time and cost are more significant as remoteness increases. If workers are paid for their travel time and reimbursed for their travel costs, then companies will look to minimize the number of trips relative to the time spent on the job site. Other factors include the quality of the transportation network available to workers and the points at which the worker becomes 'on the company clock' and needs to be paid.

Impacts and implications

The nature of the commute and the schedule of the roster rotation into and out of the work site have significant implications for both communities and workers. When we speak about community impacts, we now need to differentiate between 'home' and 'host' communities. The home community is conceptualized as the place where the worker and the worker's family live permanently – the place they think of as home. The host community, in contrast, is the location to which the worker commutes for work. The impacts and implications of LDLC are quite different for the two types of communities.

For host communities, the transition to mobile workforces raises concerns about economic leakage if the workforce no longer lives permanently in the community. If workers in the local industry spend their income on housing and other goods in a distant place of residence, there is a curtailment of local business and economic activity. This can also jeopardize the provision of local services, as some, such as education and schools are funded on a per capita basis. While many communities understand that the large workforces needed to build a new resource-development industry or facility may need to be comprised largely of mobile workers (due to the numbers and specialized trades required), most such communities want to see the operations workforce living locally. In settings where there is no adjacent or local community, LDLC may be the only option.

Inherent within the business case for the use of LDLC are considerable changes to the industry–community relationship forged during the post-war period. These changes affect both host and home communities. The economic contribution of industry to communities has become increasingly variable. Labour shedding and vertical integration processes have altered the direct economic impacts of industry (Bollman, 2007). Indirectly, industries are less beholden to the local or regional labour supply, characterized as the 'fly-over' effect, whereby communities proximate to resources are excluded from the direct and indirect economic benefits associated with exploration and extraction (Heisler and Markey, 2013).

LDLC home community impacts are less explored in the research literature. Themes investigated include spousal stress, family dislocation, substance abuse, and conflict and violence (Kinnear et al., 2013). The term 'work widow' is often used in community settings to describe, in visceral terms, the reality faced by spouses who remain at home. While stay-at-home spouses may be required to assume additional responsibilities, around domestic labour for example, there is a loss of informal family support for spouses, children, and even older parents who remain in the home communities (Newhook et al., 2011). Many of these communities may already be challenged with under-resourced services and organizations with high workloads and difficulty retaining skilled workers. Children can also be affected by labour mobility, as the absence of a parent can influence their behaviour, preparedness, and performance in school and other activities (Wray, 2012).

Worker and family opportunities and challenges

The transition to mobile work offers both opportunities and challenges to resource-industry workers and their family. Work experiences in other sectors are exposing and equipping workers with new transferable skills that will make them more resilient and better positioned to take up employment opportunities in the future. Mobile work is also providing opportunities for workers to expand or bridge their careers during economic up- and downswings across different resource sectors. During periods of economic decline in resource-dependent communities, workers have been able to commute long distances in order to maintain their income and their household in their home community.

The benefits, however, must be reconciled with the pressures that accompany mobile lifestyles. The long hours required in commuting to the work site, safety concerns associated with extended highway and winter travel, and the longer shift rotations increasingly being demanded by industry are exacerbated by the absence of fatigue-management training and access to mental health supports and other services to cope with this lifestyle. On the work site, there are suggestions of increased demands for labour productivity, given that workers have 'nothing else to do' while at the job site / camp. There also continue to be concerns about the negative impacts of mobile work on workers and their families, including household finance and maintenance pressures, restructuring of household responsibilities, concerns about family health, being away during times of family stress, more limited opportunities for community engagement, and overall mental health and well-being for both workers and all members of their family. In terms of worker mental health, The Education and Health Standing Committee (2015, Chairman's foreword) of Western Australia found that not only was the incidence rate of mental distress "higher amongst FIFO workers than in the general population" … but that there was confusion "evident around which regulator had jurisdiction for overseeing the occupational health and safety matters impacting on the FIFO worker".

There are also concerns about the cumulative impacts of LDLC over time on workers and their partners. Torkington et al. (2011) argue that isolation and changes in social environments emerge both within families and across the community, resulting in depression, anxiety, and other mental health disorders, as well as substance abuse and risky behaviours. In response, some research suggests that workers with longer shift-rotation schedules have more positively coped with family life because of reduced commuting time and costs, improved sleep patterns, and more time at home (Hanoa et al., 2011). It is also suggested that earnings acquired through LDLC enhanced the ability of households to adopt better nutritional habits, access higher quality health care, and to have more discretionary income to support family engagement in recreational activities. Community playgroups and support groups, for example, have been formed by families in home communities to share coping strategies and access broader forms of informal supports for 'functional lone-parent' families (House of Representatives, 2013).

Decline in community engagement was noted as workers and family members are unable to participate in organized sports and other community activities (Kinnear et al., 2013) and competition for scarce resources affecting interaction, cooperation, and trust within the community (Fowler and Etchegary, 2008) are also notable in the literature. As McDonald et al. (2012b, p. 24) argue, LDLC arrangements "splinter the workforce and town, generating discontent and breaking up the traditional physical intimacy of rural spaces". Shrimpton and Storey (1992) also document impacts associated with the removal of skilled workers from communities, affecting both existing operations and community capacity to embrace new economic opportunities. Despite the growth of LDLC, however, limited research explores how these experiences shape the overall capacity of home communities to respond to emerging challenges and opportunities.

Coercion–subversion within political economies

Just how mobile work fits into the processes of coercion and resistance that mark all political-economic processes and struggles has been the subject of some considerable debate. While many of the points in these debates are quite nuanced, we provide in this sub-section a simplified juxtaposition of arguments.

On the one hand, the drive towards the increasing use of mobile workers is seen not only as a significant benefit to capital, but also as a coercive force in support of Neoliberal public policy approaches and the increasing globalization of resource-industry development. The advantages to capital are quite clear and reflect an extension of the flexible labour practices and advantages that emerged beginning in the late 1970s and early 1980s. Included among these advantages are that companies only need to pay for labour when that labour is needed – and it only needs to pay for the amount of labour that is needed at any given point in time. Contracted mobile workers can be 'plugged in' to resource-development projects when and where needed and capital has no further or long-term obligations to the reproduction of that labour. In addition, the flexibility of securing mobile labour in this way also absolves capital of the costs of training a surplus labour pool for those times when specific skill sets are needed. Through binding labour into mobile work routines, capital is able to sell the notion that resource workers can live the 'good life' in urban communities supported by their temporary absences to earn income from high-wage resource-sector jobs. The collapse of the mobile labour employment market in the oil and gas sector in 2015 and 2016, however, highlighted the Achilles heel of this good life for workers – how quickly they can lose their work and be without alternatives. Such a sense that mobile is exploitive work draws upon our earlier discussions of the structural theories of capital and labour relations.

The sense that mobile work is exploitive of labour and serves only the needs of capital is, of course, contested. From a humanist perspective, it has been argued that people are autonomous actors and each seeks to maximize those things that they feel are important to them. In a world where one (generally) has to work to obtain an income, individuals who seek to maximize their income need greater choices. In resource-dependent rural and small town places, dependence upon a single industrial sector or employer can limit your choices. But if you engage in mobile work, far more options are available to you. As well, should the local employer close down or significantly reduce its workforce, mobile work allows families to stay where they might wish to live, while the mobile worker 'commutes'. Arguing the case further, mobile work may even be seen as subversive, as freeing labour from the place-bound limitations of capital. Harkening back to earlier discussions in this book, as capital has become increasingly mobile, so now too has labour been provided with an option for becoming increasingly mobile.

Conclusion

This chapter examined how the restructuring of the political economy of resource-dependent regions plays out at the individual or household level. As such, it builds upon the implications for labour described in Chapter 7 and addresses the implications of employment restructuring, job losses, and commuting on the identity, social structure, and relationships of individuals in resource-dependent rural and small town places. We also discussed the changing nature of work as demonstrated through the use of highly skilled labour, contractors, and mobile labour to address the flexible labour force needs of global capital. In terms of the broader impacts of labour restructuring on households, the emotional impacts and disruptions to children, the evolving role of women in these economies, and changes to household routines become important issues. As many places continue to experience aging-in-place, the chapter has also portrayed the implications of demographic aging for the workforce, the community, and the personal well-being for men and women.

Part IV

Change, power, and conflict in resource-dependent regions

Introduction

Part IV is about the future of resource-dependent rural and small town regions within developed economies. It is about using the political economy framework developed in this book to support a contemporary understanding of those regions as well as for interpreting the changes and transformations that they are experiencing. It is also about how that political economy framework may be used to understand the role of power and conflict in limiting, assisting, and shaping those very same changes and transformations, and in setting a platform for understanding how to turn current challenges into future opportunities.

Chapter 12 is focused on a political economy of power and conflict. The preceding chapters have identified not only the critical imperatives of a capitalist economic system, but also how a wide range of actors and institutions work and respond within that system. As change differentially impacts those institutional actors, and each seeks to both independently as well as collaboratively address their needs, the transformations themselves create and re-create ongoing contests of power and conflict. After outlining some of the core concepts central to the chapter, we then explore some of the ways by which different institutional actors are responding and the implications of these for future transformations in the global economy.

Chapter 13 takes up the challenge of re-embedding the topic of change in resource-dependent remote, rural, and small town places and regions within developed economies through our use of a critical political economy framework. The chapter explores not only the processes of change, but how they work their way through the needs and imperatives of both the capitalist economic system, as well as different institutional actors which are central to the development and re-development of these places. With attention to the historical trajectories of these changes, implications for future transformations are also drawn out, as are implications for theory – especially Staples theory and evolutionary economic geography.

Chapter 14 is organized into two parts. The first reviews the structural forces that have shaped resource-dependent communities. These structural forces are portrayed through the global capitalist economy, as well as through the restructuring of public policy and investments that have transformed the capacity of institutions and stakeholders to renew and re-position resource-based economies. The second part reconstructs the political economy of resource-based regions through our perspective as social geographers. Critical issues of colonialism, gender, aging, amenity migration, the creative class, and the restructuring of services have not only shaped social relations, but also the capacity to mobilize place-based assets and strengthen the quality of life in these places. Through this reflection of policy and community development issues, we posit a series of actions and trajectories to retool and reorganize community stakeholders and assets to be more resilient and competitive in the 21st-century global economy.

12 A political economy of power and conflict

Introduction

Change creates uncertainty, which in turn creates stress and worry. Change, uncertainty, stress, and worry are foundational to the development of conflict and contention over 'the future' and debates about who has the 'power' or authority to guide change. This chapter is focused on the political economy of conflict and power within resource-dependent rural and small town places. The preceding chapters have identified not only the critical imperatives of a capitalist economic system, but also how a wide range of actors and institutions are functioning within that system. As change differentially impacts those institutional actors, and each seeks to independently as well as collaboratively address their needs, the transformations themselves create and re-create ongoing contests of power and conflict.

The chapter is structured around four parts. The first describes the core concepts of conflict and power. The second discusses how they may play out for local residents, community groups, and local governments. The third reviews the 'shifting sands' beneath the two previously dominant players in the power hierarchy of resource-dependent places – capital and labour. The fourth shifts the scale further by focusing on the state and the global economic system – each of which is conceptually tied to locations of resource production but each of which has become increasingly distant from the particular localities of that production. A political economy approach, that allows us to include economic, social, and political perspectives and topics, creates the opportunity for a more robust portrait of contention over rural and small town restructuring.

Conflict and change

Within any community or locality, conflict and contention will exist and will, from time to time, become a dominant issue. Conflicts develop as voices rise in competition with one another. The fact that conflict exists in rural and small town places is not a new revelation (Maclean et al., 2015), but recent research from a range of theoretical perspectives has emphasized how 'change' through economic restructuring is affecting rural and small town places and how the context of 'place' is central to mediating the outcomes of those generalized pressures (Bowles and Wilson, 2016). The dynamics of small town conflict and contention are remarkable because they are so often personal (Fitchen, 1991). Within the intimate setting of the small place, conflict and debate can easily become 'attached' to individuals. Further, conflicts are not simply about immediate or transitory events. They are often grounded in past events and processes, and their outcomes then become part of the local collective memory and a foundation for the next iteration of debate or conflict.

The rural community conflict literature has its roots in the 1920s and 1930s work of rural sociology as researchers undertook a sustained intellectual effort to understand the dramatic changes transforming North America. As summarized by Sanderson and Polson (1939), community conflict:

> is 'one of the major social processes in community life … [something that] can be expected to appear when groups or individuals are making adjustments to new conditions.' Within the local community, conflict and competition may develop between individuals over differences in opinion or between groups where one stands to gain and another to lose …
>
> Sanderson and Polson [also] identify a set of 'basic drives' that stimu-late local community conflict. These drives include issues of economic or social security; adjustments to social change or to the environment; and desires to protect or promote personal or local prestige or recognition. When these are combined with the range of typical community-conflict situations, a series of flash points for local debates, competition, and conflict occur.
>
> (Halseth, 1998, pp. 47–48)

This literature adds a final important observation regarding the study of conflict over change, that being about the need for caution when trying to sort out 'true' versus 'derivative' issues in conflicts and debate. In terms of resource-industry restructuring, issues such as local job security may be substituted for other goals or the agendas of capital, the state, labour, or others.

Power and conflict – local

As resource-dependent places face current rounds of change and uncertainty, conflict is a natural counterpart to local debate. Those with vested interest in the old economy, those uncertain of their ability to make a living if that older economy disappears or changes radically, and those with financial commit-ments in terms of mortgages or loan payments can all be worried about the prospect of change. They may argue vehemently for maintenance of the status quo. In one pulp and paper town where the mill had closed, residents demonstrated this commitment to the maintenance of the status quo. In public debates, senior government officials talked about how the industry had been transforming, how the timber supply in the region might not be adequate for all of the pulp and paper mills in the region, and how there were new local and regional opportunities that they were eager to support. In those public meetings, it was clear that the community was speaking with one voice in terms of a desire to restart the old mill. When confronted with further evidence that a restarted mill would not be viable, the community again spoke with one voice and demanded the government intervene to restart the mill.

Those advocating for change also play roles in local debate and conflict. Concern over the environmental damage of resource extraction, or about the social and cultural consequences of the displacement of Aboriginal and Indigenous peoples and the near-extinguishment of their rights, may voice strong opposition to the renewal of older economic structures and push hard for new forms of economic activity. Those with innovative and entrepreneurial ideas about how to turn local resources and assets into different economic opportunities may also find themselves in conflict with those concerned about deviating from traditional paths. 'Newcomers' and new ideas may threaten the stability of traditional economic sectors.

Local governments often find themselves at the meeting point of diverging views and opinions over the future of the local community. They often find themselves being lobbied and pressured to 'take stands' or 'make statements' on one side or the other of local economic and community development debates. But local governments generally also find themselves as conflicted on such matters as those engaging in local restructuring debates. On the one hand, local governments are very dependent on their existing economic structure for the taxes, transfers, and operating revenues they need to provide local services. Mine or mill closures are not only catastrophic for workers and local businesses, but also for the fiscal health of local governments as well. As has been shown in many OECD states, replacing historically large and stable sources of local employment and revenue at the scale of small communities is not easy. At the same time, local governments are keenly aware of the range of restructuring pressures buffeting their communities, and they know the difference between the triage of short-term economic challenges and the strategic imperative of re-positioning their community and economy to be more resilient and sustainable in the 21st century. The challenge comes back to the context of local conflicts that can distract local governments from concerted and long-term attention to the strategic imperatives leaving them mired in managing short-term triage actions that do not re-position the community or economy for the long term.

Turning to questions of power and power relationships within conflict and debate about community change, we pursue the context of community power in terms of who governs and how different stakeholders benefit or lose from conflict and decision making. To assist evaluation of local conflict and debate, the question can be refined further: in terms of who gains access to decision-making power. The recasting of the community power question in this way has important implications for the geographic study of local conflict. If there are economic, social, political, cultural, geographic, or other divisions within localities, then the question as to whether some groups have been historically excluded from local power must be evaluated.

Building upon the community conflict literature, we can interpret debates about the future of a locality as part of an ongoing contest of power between different players. In most resource-dependent rural and small town places, the historical power structure deferred to capital and its local industrial incarnation. The reliance of local livelihoods and quality of life on the success of local industrial activity meant that its primacy atop the local power hierarchy was accepted and assured. As noted by Hayter (2000, p. 288) from the example of forestry towns in British Columbia, during

> the boom years of Fordism, many forest towns enjoyed prosperity and relative stability. For these communities, local development was equated with the strategies of the dominant employers typically headquartered elsewhere. Locally, the promotion of industrial development was a nonissue, and local government dealt primarily with managing basic community services.

But the erosion of industry's dominant position of power has been steady across resource-dependent places. Job losses, arguments over local taxation, withdrawals from local sourcing, and even the outright closure of local operations has meant that opportunities within the local power structure have been created. As they are created, other players are stepping forward. Individuals and civil society groups advocating various positions are using the changing local structure of decision-making power to add their voices to debates about a community's future.

Power and conflict – capital and labour

A key nexus for conflict and contention in resource-dependent rural and small town places has involved industrial capital and organized labour. In such places, both labour and capital are juxtaposed in close proximity and the workplace becomes the principal site for conflict and contention. Historically, large companies tended to dominate all aspects of community life and there is a well-documented legacy of company town labour relations. For example, Mouat (1995) describes the early struggle for unionization within BC's Kootenay mining region during the 1895 to 1901 period. With capital and labour locked in a fight over who would control the terms of work, unions organized strike actions and owners hired strike-breakers. For both sides, success had a different calculus. Owners were content with a return on their investment and this was accomplished by bringing the mines back into production. The unions achieved legislative changes to limit the use of strike-breakers.

As described in earlier chapters, within most OECD countries the participation of large industrial capital in resource-development projects after the

Second World War was generally accompanied by demands for large labour forces in order to keep the production cycle going. The organization of these large pools of labour introduced unions into the local power hierarchy. During the post-war boom decades, the local power balance between industrial capital and organized labour appeared to let both prosper. While the pendulum may have seemed to swing in favour of one side or the other over short periods of time, over the long term, industrial capital was generally very profitable and thrived, while organized labour was able to secure high wages, better benefits, and ongoing improvements in working conditions for its members.

In trying to make sense of the interplay between labour and capital, Herod (1998) argues that "social actors, whether individuals, governments, corporations, environmental groups, or, indeed, labour unions, must operate within economic, political, and cultural landscapes that may either constrain or enable their actions" (p. 1) with those actions "shaped in large part by the sociospatial context of their everyday existence and experience" (p. 23). The power of place is, therefore, central to this matrix. In resource-dependent places, the legacy of large industries and economic restructuring are part of that local geography shaping these everyday experiences.

The disruptions created by the post-1980 initiation of economic restructuring had significant impacts on this capital–labour power hierarchy. As described earlier, as capital became increasingly pressured by successive crises of profitability, it responded with dramatic transformations in its operations. While some of these responses and transformations involved investing in low-cost production regions, the ones that interest us have included the substitution of capital for labour and the implementation of 'flexibility' into labour contracts. The results have been dramatic. Flexible workplaces and job descriptions, where workers are trained to do more than one type of production job and thus can be moved throughout the plant as required, allows capital to react more quickly to changing market demands, while at the same time reducing general operating costs. As shown in many case studies, the net result of flexible job descriptions and the increased substitution of technology has been that fewer workers are required to operate the production process (Doussard and Schrock, 2015).

The intimate connection between workers and companies in resource-dependent places has made the workplace a principal site for conflict and contention. Negotiations over the terms and conditions of work mark the evolving relations between these two local groups. As debates over restructuring rolled out, 'in-situ' negotiation over flexible changes to labour contracts became contentious (Martin, 2014). The push for flexibility by capital "is interpreted by workers as a direct attack on the basic union principles of job demarcation and seniority which are there to provide dignity and stability in the workplace" (Hayter and Holmes, 2001, p. 145). Halseth et al. (2017b) describe one lengthy and contentious strike in BC's forest industry, and the even more lengthy 'shop-floor' struggle for 'community power' which followed, as an example of how the context of place and the struggle between individuals and organizations in that place for voice and power added new and human dimensions to local conflict. While the contest between a company and a union during a strike in a small forestry-dependent town can serve as a lens through which competitions for local power can be examined, it can also highlight the unevenness in capacity and access to information within those contests as well as the implications of the relative 'rootedness' of participants – with labour being more place-bound relative to the increasing mobility of capital.

For capital, decisions have made plain to resource-dependent communities that attention to profitability and shareholder interests (always the focus and imperative of the capitalist economic system) trumps attention to the needs of individual communities. For labour, the massive job reductions that have occurred have much reduced the influence of resource-industry union voices in community affairs and have increased community attention to the needs and priorities of other employment sectors – unionized or non-unionized – as local circumstances may dictate. As a result, the two historically dominant occupiers of positions of power within resource-dependent rural and small town places have each lost influence.

The outcomes of more than three decades of conflict and contention over restructuring still reverberate in these communities and within their power hierarchies. In an early study of the changing structure of community power relationships, Beckley (1996) explored changing power relationships within a small paper mill town in the northeastern United States. The withdrawal of the company from many aspects of civic life had created a power vacuum that allowed other 'players' to move into decision-making voids.

Power and conflict – state and capitalist economic system

Changing our spatial scale further, this section of the chapter pulls together an understanding of power and conflict for two of additional actors in the restructuring of resource-dependent rural and small town places – senior government and the capitalist economic system. In most of the localities considered in this book, the post-Second World War development of resource industries was a purposeful state policy. Against the need to rebuild and retool war disrupted or shattered economies, the opportunities of resource commodity exports to the new scale of industrial manufacturing capacity provided a welcome economic tool. In region after region, states supported infrastructure investment and attractive policies for international capital to set the stage for industry to invest in large resource-development projects. The success of these policy initiatives from the 1950s through to the end of the 1970s has been detailed earlier.

States not only welcome such industrial development in the post-war era, but over time they became 'addicted' (to use the word linked by Freudenburg (1992) to the benefits that industrial resource development brought – large and steady revenue flows to the state and significant employment, which created steady secondary and tertiary flows of revenue to the state. Depending upon the jurisdiction and the political orientation of the government of the day, support for industry or labour was made available through legislation and regulation. For example, both Argent (2017b) and Connelly and Nels (2017b) describe the circumstances in Australia and New Zealand respectively that supported a robust regulatory environment for labour and working people. Williston and Keller (1997), focusing on the case of forestry in British Columbia, highlighted the development of a policy environment that supported not simply the corporate side of industrial development, but specifically very large and multi-national corporate-scale investors.

That state policies to support labour or policies to support capital, all within the construct of the post-war industrial resource economy, resulted in the protection of that existing industrial resource economy is not surprising, since the state was the third key actor in what has been called the 'Fordist compromise' that "brought the interests of the state, capital, and labour in to close alignment" (Markey et al., 2012, p. 104). A significant consequence of this 'compromise' between interests is that it created a power block that was very resistant to change. Debates and localized conflicts around resource industries and contests of power between the main players were all played out within a very circumscribed arena, in which the basic structures and imperatives of the resource industry model connected to a global capitalist marketplace was not in question.

As the implications of post-1980 economic restructuring began to work through resource industries and resource-dependent regions, each of the three parties to the compromise were challenged to their core. The question in the early years of restructuring was how the particular parties would respond, which could be the most nimble and responsive, and would the triumvirate continue. The answers to these questions were reasonably quick in coming.

The first to respond, and the most successful in their response, was capital. Pressed by the imperatives of the economic system, the challenges to profitability needed to be met immediately. As noted in earlier chapters, this often put capital in direct conflict with communities, especially through losses of local jobs and reductions in local benefits. Also important were requests or demands by capital for reductions in local taxes or other operating costs. As described in the literature for many OECD resource-dependent regions, communities often had little choice but to acquiesce for such requests for fear of losing their largest employer and benefactor.

Challenges around profitability also put capital in direct conflict with the state. Following the same logic as the conflicts with communities, crises of profitability underscored requests to the state for tax or regulatory relief. Coincident with the rise of a Neoliberal public policy approach, the state was often very supportive of reducing regulatory costs and burdens. In many sectors, industrial 'self-regulation' came to be a normal policy response. As an 'addicted' party, the state was also willing to grant partial reductions in taxes, fees, and other costs in order to keep the main body of resource revenues continuing to flow into state coffers. In many cases, the state also took on, or began to subsidize, the costs and activities previously allocated to industrial interests. Over the longer term, however, ongoing reductions and changes came to mean that the state also received fewer and fewer relative benefits from resource-development activities.

As noted in Halseth (2017), the response of many state jurisdictions to long-term resource revenue declines was to try and expand the potential pool of resource-development opportunities. This expansion of state policy supports was simply a continuation of a now 60-year-old policy direction. As many observers have noted, while capital clearly recognized the new global realities of the economic system, states seemed to be grasping for older ideas and economic solutions. In a study by Halseth (2005, p. 339) of responses to the crises precipitated by the closure of the resource industries in two small towns in British Columbia, it became clear that:

> The provincial government is continuing with a 5 decades old policy to support large scale resource industries through granting use rights to large industrial capital. Despite successive government changes, the dependence of the provincial treasury on resource revenues has resulted in these governments simply reinforcing the existing structure. The recession of the early 1980s has been seen as a turning point from resource industry expansion to resource industry crisis. Despite this, provincial government policy continues to try to repair a system which has been identified as needing fundamental change.

This last point is dramatically illustrated, and 're-played', in the gold rush mentality that accompanied the 'shale gas' industry of the early and mid-2010s. Facilitated by a new technology of 'fracking', massive new deposits of natural gas were identified in historically important but currently challenged resource-exporting states such as Canada, Australia, and the United States. As in the immediate post-war period, a large resource was linked by state advocates to potentially large markets – this time the markets were in Asia. As the process unfolded and large multi-national firms staked claims to the newly discovered gas fields, debate and conflict with states arose. Under the guise of profitability concerns, capital advocated for the state to cover more and more of basic infrastructure cost for potential projects, as well as the costs of supporting training programs for skilled trades or the loosening of temporary foreign worker programs, so that any construction projects that were given the corporate 'green light' through a final investment decision could be built. As economic boom turned to economic recession during those years, capital began to demand further guaranteed reductions in future taxes, royalties, and fees. They were often granted long-term promises in these areas and on other financial matters such as how, and for how long, equipment and investment write-offs could be made against corporate taxes.

Looking forward, capital has effectively positioned itself to be more flexible and nimble in the new global economy. States have made less progress around such a transformation and continue in many cases to modify a much older playbook. Local communities are struggling as well with the economic fallout of these conflicts and changing power relations have been amplified by a Neoliberal public policy withdrawal by the state from rural community development investments.

Conclusion

As resource-dependent places continue to manoeuvre change and uncertainty, this chapter has been about the power and conflicts that complicate, and often impede, strategic actions to transform old economies. These conflicts are often rooted in the fears of how change will affect the position or stability of different stakeholders. Beginning with how power and conflict play out in local settings, we discuss labour and local government interests and pursuits to maintain the status quo of these old economy structures and processes in communities in order to preserve employment and revenues. This has made it difficult for local stakeholders to support innovative and entrepreneurial initiatives that may better position the community and local economy to become resilient and more competitive in the global economy. Moving forward, Indigenous rights will continue to challenge these positions, as they seek a greater stake in resource-development projects. In contrast, senior government policies have failed to consider new directions that may rebuild or renew mature staples-dependent rural and small town places. Instead, these policy directions have increasingly been supportive of improving the strategic competitive advantage of multi-national corporate interests, while undermining the capacity and power of communities through their Neoliberal public policy approach.

13 A political economy of resource town transition

Introduction

This chapter takes up the challenge of re-embedding the topic of change in resource-dependent remote, rural, and small town places and regions within developed economies through our use of a political economy framework. The chapter explores not only the processes of change, but how they work their way through the needs and imperatives of both the capitalist economic system, as well as different institutional actors which are central to the development and re-development of these places. With attention to the historical trajectories of these changes, implications for future transformations are also drawn out.

The chapter is organized into two sections. The first explores the processes of change. Specifically, it looks at the topics of economic change, demographic change, social change, and policy change. For each, we include an assessment of whether processes of continuity, transition, or emergence are defining the directionality of change. The second section focuses on the implications of these findings on theory. Two threads from the theory literature are picked up – staples theory and evolutionary economic geography.

Processes of change

Economic change

A globally organized economic system is nothing new, but increasing globalization means something more than just spatial scale – it also implies an increasingly interwoven and interdependent global economic and political system. Supported by institutional structures (such as the World Bank) as well as legal structures (such as international trade agreements), globalization has given primacy to capital and its freedom to move, invest, and disinvest as needed within increasingly hyper-active markets and in response to increasingly fluctuating market conditions.

While research into the processes and consequences of globalization is richly developed, there is also evidence for accelerating processes of localization. In part, this is a reaction and counter to the processes of globalization as regions, places, and people work to take some control of their place in the emergent global economic order. It is also a reaction to environmental, social, and cultural concerns around globalization and an effort to reinsert the 'local' and the 'place' into lives and economies.

In many cases, globalization and localization are occurring coincident with one another. Globalization, especially of large industry, is supported by international trade agreements and is routinely cited as evidence of the process. But we are also witnessing processes of globalization acting out through small businesses and through individuals supported by new information technologies such that goods, services, ideas, money, and labour can more freely be moved globally. The localization discussion is also present and often builds upon much older economic models, such as those supporting the growth of local businesses to meet immediate local needs, the plugging of economic leakages to grow local economies, and the shortening of transportation time and distance around access to goods and services. Some of the strongest localization debates in recent years have been around matters of local food production.

If the core action of globalization is to support a more homogenous and seamless economic, social, political, and perhaps even cultural (when money and market exchange are key cultural attributes) playing field for capital, then the mobilization of capital is central to the project. Such mobility is supported by various forms of international trade institutions, technologies, and agreements. But there are consequences that come with any support structure. For example, one of the consequences of international trade agreements is that they generally place limits on the abilities of individual states to intervene in their own economies. They also typically provide venues through which capital can request compensation for changes in state policy, whether they had significant investments in that state's jurisdiction or not.

The mobility of labour is another big issue in globalization and international trade agreements. As such, they are often linked to the various forms of temporary foreign worker programs that operate in, and between, different jurisdictions. These can be important in circumstances where the operations of capital in high-cost jurisdictions seek to reduce their overall costs of production through the use of cheaper labour. A recent chapter by Argent (2017a) certainly highlights the importance of different temporary foreign worker programs in the agricultural sector in Australia. However, as we also saw earlier in this book, labour is not simply a pawn in the globalization of labour process, it has actively turned this new operating environment to its own advantage. This can be seen in the mobility of highly skilled labour within developed economies as people move to apply their talent wherever there are opportunities around the world. It can also be seen in the growth of consulting firms, telecommuters, and others who are able to 'live local and work global'.

The mobility of labour has significant costs for both the commuting worker and for the home and host communities connected to such mobile labour flows. Opportunities for the worker to earn income, learn new skills, and grow their contacts to assist with securing future work are countered by the physical and mental health costs of long-distance labour commuting (Ryser et al., 2016b). Thus far, these significant costs have not been fully recognized or addressed by public policy. Depending upon the circumstances, however, these costs may have been recognized and addressed by individual economic sectors or by individual companies within sectors. For home communities, the income that mobile workers bring can secure and contribute to the local economy, while at the same time the functional 'loss' of a talented and skilled part of the local population has follow-on impacts on families, voluntary groups, and sport / recreational / cultural community organizations and activities. In host communities, the assistance that skilled workers offer to support local economic activity come with costs that include the leakage of local income wealth to outside regions and the additional costs of services and supports for the large numbers of mobile workers when they are 'in' the work region.

All of these issues come together in terms of creating, in resource-dependent rural and small town places, a new landscape of precarious economic relations. In this case, everything is increasingly based on contractual relationships. Historically, people might secure jobs with a firm and be employed in those jobs for their entire career. Increasingly, however, jobs, sales of products, provision of services – everything – is being contracted out. Industry does it. Business does it. It turns out that labour is doing it as well. The state is also doing it. The net result is that long-term functional economic-employment relationships are being replaced with precarious positioning in the labour force and society. The 2015 downturn in global energy prices demonstrated this precariousness in the oil and gas sector, as workers around the world (office and field-based) who had enjoyed years of ongoing contract renewal and employment were let go. This highlights how that precariousness of employment is exacerbated by the globalization of the economy and the ability of capital to respond quickly to change and threats to its profitability or return to shareholders.

By re-embedding economic change in resource-dependent regions, it becomes clear that there is a great deal of 'continuity' in activities and processes. Many large industries are stuck in their existing commodity 'boxes'. They have significant investments either in the resource or in their processing facilities such that they struggle hard to stay competitive in the commodity marketplace. States, it seems, are also stuck in the resource commodity box. They are very dependent on the resource economy and to the revenue streams to which it contributes. Their regulatory and governance framework understands those resource commodity economies and they are reticent to change, even though warning signs about the need to change are now over 40 years old. We also find that many resource-dependent communities remain stuck in the commodity trap. This is the economy and community that they know – the seasonal rhythms, the jobs, etc. The very definition of the community itself, and its residents, is often rooted in these traditional economic sectors. For all these players, the pace of the global economy and the well-known challenges of a staples economy make continuity a perilous pathway.

Turning to processes of 'transition', we are seeing examples of how this is underway in a number of the traditional commodity sectors. One of the most common examples used today is the bio-economy. In the case of forestry, traditional dimension lumber or pulp and paper producers in the sector are now thinking of allied products or entirely new product streams through adoption and development of the bio-economy. Within North America, these transitions are just getting underway in small ways with some individual companies. In Europe, more progress is being made on the bio-economy front – for example, with the retooling of former pulp and paper mills into bio-refineries. Some of this difference might be explained by the significant costs of developing new processes and products, and how such is being assisted by funding support through the European Union.

There is also a good deal of evidence in support of processes of 'emergence' in the economy of rural and small town resource-dependent places. Next rural economy developments that take advantage of the unique assets of localities, that build upon a place-based approach, and that actively work to re-imagine and re-bundle those assets in ways that fit with local aspirations are creating new opportunities within local, regional, national, and even global market-places. They are part of efforts to diversify local economies by adding new economic sectors, bridging between economic sectors, and searching for new opportunity within established economic sectors. They are also looking at ways to monetize components of local quality of life and amenities that are available in order to create new economic potential. Efforts at emergence are found in the work of industrial, community, and state interests. Together, they seem to have actively incorporated an understanding of the political economy of change in resource-dependent industries so as to build more robust and diverse economic foundations (for their business, their jurisdiction, or their residents) for both sustainability and resilience.

Demographic change

One of the most significant challenges facing resource-dependent rural and small town places is population aging. Population aging is a national trend found within all OECD countries and so the coincidence of national trends with the specifics of resource frontier aging creates a unique challenge. As described earlier in this book, the specifics of resource frontier population aging is the result of a timeline that began with the creation of significant new resource industries in the rural and remote hinterland regions of developed economies after the Second World War. Over the long post-war boom, these workforces remained relatively stable and began to age-in-place. With the coming of intense economic restructuring, the presence of unionized workforces and labour contracts that favoured rules around seniority, it tended to be the more recently hired younger workers who were let go as operations downsized. This exacerbated the trend towards a generally older workforce.

As the older workforce now found in many resource-dependent places and regions begins to retire in large numbers, there are significant implications for industries, states, and communities. One of the first issues for industry is how to transition an older workforce via succession planning. What roles would mentoring and workplace shadowing play in succession planning? These implications also play out for businesses as well in terms of exploring ways to take advantage of new employment opportunities, to cope with worker shortages, and even to attract new entrepreneurs through succession planning for their own retirement. Population aging also has significant impacts for the local economy and the local demography because in many OECD countries the state has yet to really grapple with what it will mean to have large numbers of older residents in these regions and communities. These are regions and communities that have not typically had large numbers of older residents, with the result that they lack investment in needed services and infrastructure. Further, as population aging works through the urban population as well, cash-strapped state agencies will find it difficult to address needs in rural and small town places when the urban-based media directs regular attention to the need for such investments in large cities.

Linking the economic and the demographic in a simple way, the retention of retirees in rural and small town places is a vital part of local economic planning. Having built their retirement savings by working in the resource economies of these places, it is important to hold these residents locally when they retire so that they spend their retirement savings within the local economy. In this regard, seniors are a form of basic sector economic activity that will bring money from outside the community (pension income) to be spent locally. However, most states have been lagging in terms of investments in services and supports to retain seniors in rural and small town places. Needed investments, for example, in the health field include transitioning from a medical model to a preventative model, so as to support the ways and services that will help keep people healthy and happy in their own homes for longer. In terms of housing, this includes not only adaptation of the existing housing stock, so that people can remain living safely and longer in their own home by undertaking various forms of accessibility renovations, but it also includes providing new housing options for when people wish to downsize and move through a continuum of housing towards care.

The voluntary sector in rural and small town places is also impacted by population aging. Older cohorts of volunteers need to be replaced and this also speaks to the need to pay attention to succession planning and the need for building capacity among younger people to participate. Population aging also increases the risk of volunteer burnout (through older volunteers becoming tired or through an increase in the demand for volunteer-delivered services to aging households) in communities where there simply are not large numbers of people who would be able to step up as potential volunteers. In some ways, population aging issues also connect with the debates about temporary foreign workers and local labour mobility issues, as we find the globalization of workforces being cited as one opportunity for backfilling labour requirements. This line of thought also fits well with the precariousness of economic relations identified earlier.

When we consider the many pressures that arise from demographic change, we also need to look forward and think of the next generation workforce. Generally, there has been a failure to recognize the need to reinvest in rural and small town places so that they are attractive (through 21st-century-appropriate services, amenities, housing styles, communication infrastructure, etc.) to the next generation of workers. There has also been a failure to recognize the different forms of educational and learning processes that this next generation workforce will require in order to be successful in their transition into, and movement through, working opportunities. There are also some challenges in that many of our educational and training programming remain geared to an older economy framework and there are fewer opportunities for young people to become equipped for the different economic futures of the contemporary global economy.

The next generation workforce, as a result of its smaller size, compared to the baby boomers they will be replacing in the labour force, is also going to be a workforce that is able to choose where it works. This competition for labour will disadvantage rural and small town places if they do not invest in the kinds of quality-of-life services (i.e. housing and infrastructure) that this population cohort is interested in. Bringing together issues around demography, we see continuity in terms of older labour force practices persisting, whether that is in the structure of management, labour, union organizations, training practices, etc.

We also see some transition occurring with respect to a readiness for demographic change. To start, many rural and small town places are recognizing the importance of retaining seniors and are actively working to support communities suited to people of all ages and stages of life. State investment, recognizing the impossibility of all seniors moving to metropolitan centres, are also starting to transition supports and investments to the retention of seniors in smaller places. There is also transition in training opportunities and the way in which life-long learning will be a bigger and bigger part of people's working lives.

The emergence of a next generation workforce, with its own unique goals and aspirations for fitting into the next generation economic activities being created through 21st-century technology and society, is the key to demographic re-balancing in rural and small town places. As noted, these places can compete very well for younger workers and households if suitable investments are made to support the notion of living locally and living globally.

Social change

Social change is a large topic area. As long noted by sociologists, societies are generally always 'emergent'; that is, they are always in a process of adaptation and change. For our interests here, we select and highlight a couple of macro themes in social change. One of note involves the growth in environmentalism and environmental awareness (Sherval and Hardiman, 2014). At the same time, there has been increased recognition of human and Aboriginal / Indigenous rights. These have had some important impacts for resource industries in developed economies. Increased environmental awareness has led to increased regulation of industrial activities, protests around new industrial activities or sectors have restricted or directed investments, the designation of some undeveloped areas as parks instead of for resource development has limited the potential land base for industrial resource investments, and industry–community–state–environmental group accords have defined new working relationships in some very large resource regions. Where resource industrial practices continue, market-focused protests have led to the creation of responses such as environmental certification regimes (Hackett, 2013), so that consumers can take comfort in the fact the products they purchase were sustainably harvested.

We have also seen new ways in which that environmental awareness or recognition of human and Aboriginal / Indigenous rights seem to have been contradicted by individual and social actions. Increasing environmental awareness, for example, seems to also come with examples of increasing environmental waste and degradation (the rapid expiration and near semi-annual replacement of personal electronic and computing equipment comes to mind). Similarly, the imperatives of immediate jobs and economic activity often override our sensibilities around longer-term respect for the environment or rights.

There is also transition in society about how we value work versus how we value the quality of our lives. In the years immediately after the Second World War, people were willing to sacrifice their work environment in order to provide stable housing as well as a stable context for their families and for their children. This mentality was built upon the experiences of decades of upset and uncertainty which the depression and the war had brought. Now, attention to self-fulfilment and immediate personal gratification often guides decision making beyond the ethic of working for a living. This impacts both economic and community interests.

There are generational differences emerging in other areas as well, including with respect to ways of knowing, ways of learning, and ways of communicating. The growth of new information technologies has played a significant role here. Taken together, some of these economic, demographic, and social changes mean that people are now more willing and able to choose to live in particular localities and work in other regions or localities. This ability to live anywhere and work anywhere will challenge resource-dependent rural and small town places if they do not invest in 21st-century amenities, services, housing, etc. in order to attract and retain those workers and their families. They have to create places where people not only want to create careers, but also want to grow their families and create their lives. This is the same challenge faced by resource-town planners in the 1950s; while the problem is similar, the solutions will be different, but the need to invest remains the same. As noted earlier in the book, rural and small town places generally compete very well for the aspirations of many next generation workers, but these places need investments in order to fulfil that potential successful competitiveness.

We also find that as society is changing, some of the key institutional actors are not. Government and civil society, as two examples, are too often still using generations-old, if not centuries-old, structures and practices, in terms of their organization and how the ways participation in those organizations is managed. What we have learned about new information technology and next generation workers is that those forms of organization and participation do not necessarily fit with how people are now living their lives. We might think, for example, of a local voluntary organization structured within a club or society framework holding weekly meetings managed by a very hierarchical leadership structure in order to deliver its service or to meet a specific need. In contrast, we might think of a similar need that is met within a short matter of hours by a group of people who organize a flash mob to bring funding support to a family who has lost their home through a tragic fire. In the first case, those 20th-century weekly meetings and bureaucratic structures might appear dull and boring compared to the exciting 21st-century world of social-media-organized relationships and technology as a way of engaging with society and society's needs. At the same time, we see that mobile labour and labour mobility supports a further potential disengagement with local civil society. Not having the time, and not being local for long periods of time, means that people are just not able to participate in those 20th-century models of voluntary-sector societies and organizations the way that people did during the long post-war boom. Transforming groups, organizations, and institutions into 21st-century models in terms of structure, organization, and participation will be important to the viability and flexibility of rural and small town places.

A last important area of social change that we wish to note has to do with the changing nature and structure of households. For decades after the Second World War, the development of most resource-dependent places and regions was based on the model of the nuclear family. Progression within the model involved children growing up in two-parent households, who upon graduating from high school then went off and formed their own nuclear households and families. Today, not only is the structure of parenting very different, but so too are the timing and processes whereby children move out of the house. Children may now move home sometimes permanently, sometimes on a rotating basis during periods of transformation in their lives or in their participation in the economy. There is also increased discussion of the 'sandwich generation', as older households are not simply continuing to have responsibility for their grown children, but are also taking in their senior parents, either by choice or when the state may not be providing sufficient supports. These are just some of the ways by which households are coping with economic change and uncertainty for all members.

The question of continuity is a difficult one for the topic of social change. As noted above, society is generally understood as experiencing ongoing processes of transformation. Thus, it is right to ask: Where might there be continuity? At its most basic, we find continuity in the way people need money in order to secure housing, to feed themselves, to support their family, and to pursue quality in their day-to-day lives. For most people, that money is earned through work. As Halseth (1999) noted, resource-dependent places and regions are those where people went for the opportunity to work and where they stayed for the quality of life that such settings offered to them and their families. The need for income and money continues and this is one of the push-and-pull factors in resource-community change. Those engaged in traditional economic sectors often push for a continuity within that sector. After all, the world still needs the resources and there are still opportunities for industries and for workers. There are also those trying to pull the local economy onto a more diverse footing by adding new economic sectors that can capitalize on new 21st-century opportunities.

Those trying to diversify the local and regional economy mark evidence of transition in rural and small town social structures. They introduce new people, ideas, habits, practices, sensibilities, priorities, and opinions to the local mix. In turn, local government and civil society actors are also transitioning. The example of new information technologies being adopted by both illustrates the ongoing transition and re-positioning of institutional structures within these communities. Similarly, transition within the business and industrial workplace reflect changes in society generally and changes within the social construction of resource-dependent rural and small town places in particular. A social geography approach tells us that those relationships shift over time and are constructed and re-constructed in place. Issues of gender, work, power, 'race', and ethnicity are all transitioning under the influence of local through to global processes which are shaping social change.

Taken together with the opportunities of demographic aging, processes of transition within places and communities also blend into processes of emergence. The replacement of a space-based economic development approach with place-based understandings of the local economy and community can lead to the emergence of new opportunities and directions. Significant amongst the emergent processes within resource-dependent rural and small town places is a growing re-awakening, or awareness, of Aboriginal / Indigenous rights. In the post-war development era, jurisdiction after jurisdiction too often ignored those rights in the drive to develop new resource industries. Today, emergent forms of reconciliation and collaboration are trying to rebuild social relationships at the corporate, government, community, and individual levels. There is recognition that not only diversity in the local economy, but also diversity in the local community, are key contributors to resilience and sustainability. While many resource-dependent regions were socially and culturally diverse as a result of the post-war immigration for work, there are new and emergent aspects to that diversity.

Policy change

The last topic we wish to pursue with respect to re-embedding processes of change into resource-dependent remote, rural, and small town places and regions concerns policy change. The macro story in this section is one where public policy in many regards appears to have lost its way. After the Second World War, the opportunities of the global economy, the abundance of natural resources in many OECD countries, and the clear guidance of a Keynesian public policy framework created robust and cross-government strategies, such that many states were able to organize a concerted transformation of their economy through directed public policy.

In the intervening decades, of course, we have described the transition from a Keynesian to a Neoliberal public policy approach. We have also described many of the inherent contradictions of that Neoliberal public policy approach relative to the intermittent need for states to re-adopt Keynesian interventions during times of economic uncertainty or corporate irresponsibility. In many respects, today's public policy approach in many states is something we describe in Chapter 6 as a 'reactionary' era of both public and corporate policy approaches that countenance limited state regulation and intervention, but in a context where the state has been hobbled by internal reductions in capacity, so that it is often unable to adequately manage those regulations and interventions. Some of the implications for transforming rural and small town places within resource-dependent regions are that the state has inadvertently, though sometimes purposefully, reduced its own capacity to respond by downsizing its regional staff and offices in response to calls for 'smaller government' or reduced taxation. This reduced capacity means that the state has lost some of its responsiveness in the face of rapid change and it has lost some of its capacity to re-imagine the future and get ahead of global economic and social trends with pro-active and integrated policies. It has lost many of the tools that it had for monitoring change on the ground in resource development regions that could act as early-warning systems, and this, when coupled with international trade agreements, further hinders and limits the state's capacity to respond to transformation.

The state has also downloaded some of its former activities onto local governments or onto the voluntary sector. As noted earlier, it has done this without downloading sufficient decision-making authority or fiscal resources to those same bodies. The net result is that many local governments and voluntary-sector groups have become overburdened with activities and stretched and strained so that even when the state wishes to be a partner in community development and transformation in these settings, the locality often is hard-pressed to participate.

A further challenge to the Neoliberal dictate of offloading state activities has been that even when financial or policy support is offered in service areas, it is now accompanied by a significant, if not crippling, demand for oversight and accountability. Small organizations, already burdened with downloaded or offloaded tasks, may receive small levels of funding support, but that funding support comes tied with significant demands around paperwork, management, reporting, and administration. These very much speak to a continuing fear that the state has with the activities of local governments and voluntary-sector groups and the need to regulate them through management and administration.

A third critical area of change as a result of policy shifts concerns the rotating processes of industrial de-regulation and then re-regulation. These are often 'knee-jerk' in response to either crises of profitability or where de-regulation is argued as being necessary to free industry to become more efficient and reduce its operating costs, or re-regulation in times when something has gone catastrophically wrong with an industry or sector and the public demand oversight. Such changes can also come as a result of the changing ideology of governing parties who 'believe' in regulation or deregulation as absolutes. Again, these transitions and reversals speak to a public policy framework and structure that is losing its concerted ability for renewing the community development vision for resource-dependent rural and small places in the 21st century and is often unable to stay the course with concerted and cross-government coordination of activities to support that vision.

A fourth area of challenge is that while economies and resource-dependent communities are changing, access to information and participation in the debates about those economic transformations can at times be difficult. This is especially the case where development debates around new economic opportunities involve formal hearing or assessment processes. Those processes have become highly legalized and thus circumscribed to the point where participants guard information and may even seek to limit participation in the process (Gillingham et al., 2016). In some cases, it can become difficult for all but the most dedicated and most educated to participate.

Some of the consequences of these generalized trends in public policy change have led jurisdictions in many OECD countries towards a disjointed approach to public policy making. Sometimes the habit of responding to sensationalist news or public opinion trends reinforces that there is a lack of concerted policy or development direction. Second, sometimes old solutions are often rolled out in response to newly constructed challenges. One of the often-cited examples is with workforce transition where educational programming frequently rolls out older models such as apprenticeships in order to deal with shortages in areas such as skilled trades, while failing to recognize new and next generation economic opportunities that also will need support and structure for the training of workers. The result can be a piecemeal public policy approach that lacks coordination, lacks cross-government vision and engagement, and lacks an implementation framework that understands the long-term benefits of investment. Instead, the response too often is a series of one-off policy triage.

Summarizing for policy change, we see continuity in terms of a failure to build significant wealth and reserves after decades of resource extraction.

Among developed economies, Norway stands as one example of a state that has taken deliberate and significant steps to using resource wealth to create a fund that can support its economy through periods of resource uncertainty. There is also continuity in terms of a more general failure to build a public policy framework that can adequately see the state through these transitions. States too often seem to be stuck with older models and few have taken the needed steps towards retooling the structures of governing, and the associated institutions and policy supports, to fit the new economy.

There are some areas of policy emergence. Most notably evidence in this more 'reactionary' era comes in the form of attempts to incorporate almost 'neo-Keynesian' supports and approaches into the more open market-based framework of Neoliberalism. Just as rural and small town places need to retool to be responsive and resilient against the realities of the 21st-century global economy, so too do public policy frameworks need draw upon recent lessons in order to retool policy approaches to these same realities.

Implications for theory

This second section of the chapter focuses on the implications of these discussions on theory – especially staples theory and evolutionary economic geography.

Staples theory

As one of the longest-used theoretical frameworks for understanding resource-dependent economies, what are the implications of our political economy exploration for Staples theory? Before starting, we recognize that Staples theory itself emerged from an explicit political economic framework. It highlighted not only economic processes, but also the social and political processes which brought to the fore the role of institutions and institutional structures in creating and perpetuating staples economies.

This exploration reinforces that a Staples theory framework continues to be valuable in understanding and interpreting change and continuity in resource-dependent rural and small town places and regions. As first argued by Innis, the structural forces of geography, technology, and institutions continue to play important roles. The global economy persists, and the global economy continues to demand natural resources as inputs into advanced manufacturing processes. Expectations of increasing growth in the mean wealth of such massive markets as those found in India and China will continue to push demands for consumer goods into the future and, thus, it is expected that staples production will continue to be a critical component in the economies of resource-dependent regions and a critical feeder into the global economy. That said, of course, the global economy is dramatically impacted by processes associated with globalization and these processes (assisted by formal trade agreements and various information technologies and innovations) mean that the pace of the global economy is accelerating. This acceleration makes it more responsive to signals in the market and associated economic up- and downswings. While the globalized economy may be more responsive, resource-producing regions and the institutional structures on which they depend are not yet as nimble and responsive.

One of the challenges for resource-dependent places and regions is that long-standing political structures still persist. These are the institutional frameworks for managing access to staples. But those political institutional frameworks are being challenged by globalization. Key in this challenge is that most governing and policy structures cannot react as quickly as the world markets to opportunities and challenges. In addition, there are also institutions associated with labour, associated with civil society, and associated with civic society that are important players in the continuing development of resource exporting economies. All of these are also challenged by the accelerations brought about by globalization. The result is that they will continue to be buffeted in more dramatic and more extreme ways. Rather than being pro-active with new visions and long-term investments to re-position resource-dependent regions for future competitiveness and community / economic diversity, existing political structures and institutional frameworks too often are, at best, reactive and well behind global market trends – and too often when they do respond they are trying to replay 1950s industrial resource development policy in a world that is today very different with respect to attitudes around development, resources, the environment, cultural rights, etc. The self-reinforcing rigidities in staples systems can often support continuity in economic orientation and approach – with the result that the challenges of dependency and truncated development can also persist.

It is particularly important to note the changing role of civil society. Across OECD countries, civil society, whether rooted in resource-dependent regions or in metropolitan areas, has been successful in moving into the decision-making power structure and placing its voice in the discussion and debates around resource development. This will continue into the future and be an important part in how staples commodities are understood and potentially mobilized into economic production.

That said, the central limitations and challenges to a staples economy that were identified years ago remain. Staples economies continue to be vulnerable to the demands and prices generated in the global economy. As noted above, the processes of globalization have accelerated this vulnerability and made long-term investments in such industries more precarious. Also persisting are the processes around truncated development as international capital invests in particular resource-development regions for particular reasons relative to its management to a global supply and marketing network. This continues to put a drag on the need for more, and more concerted, attention to ongoing economic diversification in rural and small town places and regions.

While resources remain rooted in place, capital, and in some cases labour, is increasingly mobile and this adds to the volatility in resource-dependent regions. This transformation from the model of a fixed resource town located next to a large mine or mill, to a much more fluid mobile environment where both capital and labour are choosing where to live, work, and invest, means that attention to investment in non-metropolitan regions to support quality of life and attract not only economic investment, but also the workers and their families is even more important than at any time in the past.

Finally, of course, all of these trends suggest that many rural and small town resource-dependent regions are struggling with the notion of path dependence. Resource commodity exports remain a viable option for the future but they come with significant and well-known challenges. How then to best manage the opportunity if continued resource-sector activity with those vulnerabilities and challenges. This makes a natural linkage to the dialogue and debate within evolutionary economic geography.

Evolutionary economic geography

The central debate in evolutionary economic geography is about whether path dependence persists or whether there is a potential for incremental, or evolutionary, change. Through this review of resource-dependent rural and small town places, we have found evidence of both path dependence and change. Our exploration, as well as the cases covered in a recent international volume (Halseth, 2017), identified a number of national contexts in which continuity within the old economy structures and institutions, transition in some of those structures and institutions, and even the emergence of some new economic and institutional players all exist coincidentally within places and regions. Some of the challenging issues that confront or constrain processes of change include that power hierarchies are deeply embedded within the institutions that are themselves embedded within these regions. This structure of power hierarchies is very resistant to change. Further, on the industrial side, the clear imperatives of capitalism for profitability and return on investment to shareholders constrain the flexibility of industry at different times in the economic cycle. This also has seemed to put a drag on innovation and change. For the institutions of civic society and civil society in rural and small town places, there is also the very real challenge of managing short-term triage activities to the recurring challenges / opportunities of economic upswings and downswings relative to what is known about the need for long-term focused attention on strategic transformation to build for a more sustainable community and resilient economy.

Within evolutionary economic geography, the construct of regional path dependence is impacted by the roles of technological innovation, bounded rationality among economic agents, and the role of institutions in driving economic systems. These are reinforced by habit, by the sets of embedded routines found within the day-to-day activities firms, policy making, and in the practices of institutions. Across each sector, our exploration has found limitations to transformation at the same time that it has found long-term imperatives that should have been motivating for, and driving, such transformation. Evolutionary economic geography has emphasized the twin processes of continuation and change, with transformation in social-political-technological arrangements first replacing, and then becoming embedded within, temporally bounded regional economies. As we have seen, the contingencies of place continue to matter, as does the interplay between global capital and local interests.

Finally, the commodification of land further impedes the ability to transform and undertake innovative and creative responses. While place-based development may talk about the need to re-imagine and re-bundle assets in creative ways to meet new economic opportunities, a recent volume on cumulative impacts identified how the allocation of lease or tenure rights to vast tracks of land for the extraction of specific natural resources pre-determines the economic fate of whole regions for decades if not generations to come (Gillingham et al., 2016). Linking past activities and decisions with opportunities for the future, tracing the cumulative impacts of those past activities and decisions and combining them with the expected cumulative impacts of any added activities, is challenging. It is not readily handled in the assessment processes in use today (Uhlmann et al., 2014).

That there are signs of evolution or emergence in resource-dependent regions becomes somewhat muted if power hierarchies, institutional structures, and broader policy approaches do not change to cement this trajectory as stakeholders become fixated on immediate pressures, the outcome of which otherwise becomes piecemeal and ineffective for broader transformation. We build upon these notions to suggest that institutional changes should also be informed by place and place-based assets.

Conclusion

This chapter on the political economy of resource town transition has detailed economic, demographic, social, and policy change through processes of continuity, transition, and emergence in resource-dependent regions. This embedded change is increasingly grounded in contractual relationships and fluidity of capital, assets, and labour that have intensified the precariousness and uncertainty for these places in the global economy. While local, state, and industry responses have often remained stuck in old economy imperatives, changes have also been concomitantly supporting processes of transition and emergence of place-based assets, new sectors, and possible pathways. In some cases, these target the necessary transformations that are needed to address the needs of the next generation workforce and older workers who wish to retire in resource-dependent rural and small town places. In other cases, recognition of Indigenous rights and environmentalism are playing important roles to rebuild relationships across local, state, and multi-national interests. More often, however, Neoliberal public policies have reduced or limited the capacity of places to support the types of investments and changes needed to better position these communities in the new global economy. In the latter section of this chapter, we have situated these lessons within some theoretical implications for Staples theory and evolutionary economic geography. At the heart of these implications is the underdeveloped responsiveness of institutional policies and structures to react to rapidly changing market conditions that, unfortunately in many cases, continue to reinforce dependent pathways in resource-dependent regions.

14 Development embedded in places

Introduction

In constructing a political economy of restructuring and change in resource-dependent rural and small town places and regions, we have approached the question from the perspective of social geographers. While other writers will emphasize other topics, especially economic topics, we have focused on the socially constructed and relational aspects of the economies and societies so profoundly impacted by increasing globalization. This closing chapter brings together the central themes of the book. It addresses the ways by which we are able to understand the political economy of rural and resource-dependent regions within developed economies as well as the historical and contemporary transformations they are experiencing. Against this context, the chapter is forward-looking in terms of describing the potential future for our subject remote, rural, and small town resource-dependent places.

Whenever we write about issues like regional and community development, it is important to recognize that we are writing about long-term processes of change. While we can recognize these long-term processes, and hypothesize about the future implications of their trajectories, it is always important to note that we are still writing at a moment in time. This may give primacy to immediate issues, trends, and debates even though we struggle against these tendencies. With these caveats, we try to both draw upon the imperatives and trajectories described in the book and also forward to speculate on issues in the near future for resource-dependent rural and small town places in developed economies.

The chapter is organized into two parts. The first reviews the structural forces that have shaped resource-dependent communities. These structural forces are portrayed through the global capitalist economy, as well as through the restructuring of public policy and investments that have transformed the capacity of institutions and stakeholders to renew and re-position resource-based economies. The second part reconstructs the political economy of resource-based regions through our perspective as social geographers. Critical issues of colonialism, gender, aging, amenity migration, the creative class, and the restructuring of services have not only shaped social relations, but also the capacity to mobilize place-based assets and strengthen the quality of life in these places. Through this reflection of policy and community development issues, we posit a series of actions and trajectories to retool and reorganize community stakeholders and assets to be more resilient and competitive in the 21st-century global economy.

Structures and actors

In many ways, the chapters of this book have highlighted two sets of forces acting upon resource-dependent rural and small town places. The first involves structural forces associated with the capitalist economy, the increasingly globalized economy, public policy transitions, etc. The second set involve the legacy of decisions and investments that have occurred over time and in place as people, industries, and states sought to negotiate and re-negotiate opportunities and challenges within those structural forces. For the communities and regions that we have examined, place is the critical variable as localities and regions become the sites for the collisions of these two sets of long-term forces.

Capitalist economy and framework

As noted early the book, we live and work within a capitalist economy. While there are many descriptions and analyses of this economy, we have drawn specifically upon a Marxist analysis so as to highlight its important structural aspects. While the visible structures and manifestations of the capitalist economy seem to have changed a great deal over the past 70 years, our review of the nature of capitalism highlights that its underlying organization and imperatives remain relatively unchanged. The shift from goods to services, relocations of manufacturing activity to low-cost production regions, and the ebb and flow of industrial giants from manufacturing into technology firms, are all part of the changes seen in the surficial expression of capital. Beneath this, it has been noted that the imperative to generate profit, and a return to capital on investment, remains unchanged. To support this imperative, the need to subordinate labour, and to a degree the state, to the ability of capital to create and re-create wealth also remains. Debates and conflicts on many topics and issues reflect the ongoing contestation of the capitalist economy, as well as the ongoing adjustments being made by capital, labour, the holders of other key policy or resource factors, as well as the resource-dependent places and regions of developed economies.

For rural and small town resource-dependent places within developed economies, the continuing imperatives of capitalism have had important implications. In many long-standing resource-producing areas, especially those focused on mining and forestry, the cost–profit squeeze on capital has underwritten extensive job losses and closures. Where such industrial activities persist, local benefits are reduced and uncertainty is the watchword. For some resource-producing areas, especially those tied to oil and natural gas, there have been dramatic booms and busts in recent decades. The expansion of fracking technology has certainly changed the production landscape in more than one way. In many cases, there may be significant new job opportunities, but the challenge is whether the workers will live locally (and thus transfer the benefits from resource development into the local economy) or whether they will participate in the increasingly common pattern of long-distance labour commuting.

One of the key places within which the ongoing contestation of the capitalist economy is taking place is within public policy. The transitions from a Keynesian to a Neoliberal public policy approach, and the modifications of an almost 'neo-Keynesian' approach when the contradictions of a Neoliberal framework become too apparent, have traced the ebbs and flows of this contestation. For the communities included in this book, the impacts of this more 'reactionary' era in public policy approaches (see Chapter 6) have been significant. With greater freedom to invest and mobilize capital, we have recounted cases when rural and small town regions have either lost economic and employment opportunities or gained economic and employment opportunities. These changes, and where and when they occur, tend to be very dramatic. As places within developed economies, one difference in the community and economic development task is that new economic opportunities tend to come forward with much reduced employment opportunities as all resource-industry sectors struggle to compete with low-wage, low-regulation regions across the globe.

For rural and small town resource-dependent areas, the Keynesian to Neoliberal public policy transition has had similarly negative implications. Senior government participation in maintaining key infrastructure and services has been scaled back or withdrawn altogether. Service and infrastructure deficits are the language across most OECD resource-dependent regions. Senior governments have also lost their 'eyes and ears' on the ground such that centralized policy and program decision making is less well informed about what is actually happening in those regions that are so central to the economic health of the state. As opportunities and challenges arise, local governments are increasingly on their own. As the demands for assistance to laid-off workers, assistance with accommodating large numbers of new workers, supports for languishing and outdated services and amenities, and other pressures grow, local governments are very much stuck with a jurisdictional mandate and a set of fiscal resources that have not changed significantly since the 1960s.

Linking to the surficial-underlying components of the capitalist economy are those found in the global economy. The heated debates about globalization certainly note the increasing flexibility and mobility of capital and the increasingly interconnected policy environments that global trade and bi-/multi-lateral trade agreements commit states to following. But these speeded-up manifestations reflect only the current iteration of globalized trade and global trading company arrangements that can be traced back hundreds of years. Even while debates within individual states may swing from 'globalization' to 'protectionism', and back, there can be no doubt that the interconnected global economy assisted by high-speed communication and transportation will persist and that such transitions in either policy or practice will simply continue to be part of the way the entire global economic system shifts over time.

For our resource-dependent rural and small town places, there is no question that processes of globalization are important. They impact speed and direction of economic change, they impact the capacity of state intervention and support, and they impact the mobility of both capital and labour. These impacts represent both opportunities and challenges. They also represent an imperative for rural and small town places to be informed about their embeddedness in this framework of structural forces if they are to make better choices about their own future. As we often say to local leaders in our community development work, 'this is not 1970, it is not going to be 1970 again, your regional economy is never going to look like its 1970 again, and so we need to make 21st-century choices'. Looking forward, it is clear that resource-dependent places and regions must work with an understanding of the realities of the contemporary capitalist economy and its global tendencies. They must also work with a realistic sense of their own position within their respective regional, national, and international contexts. That said, opportunities for places and regions to re-imagine their natural and community assets, re-bundle those assets in new ways to create new opportunities, new competitive advantage, and to meet local goals and aspirations, remains the best path for building economic diversification and community resilience. Place-based community and economic development in an increasingly globalized world continues to make good sense.

Trajectories

As we have noted in many places throughout the book, the choices and investments made in the past create a legacy of impacts, give direction, and create drags on our ability to choose and invest now and into the future. To start, resource extractive activities and economies have been part of the non-urban landscape for a very, very long time. The value to be gained by trading a locally abundant resource or good to other places in exchange for a resource or good not locally available has also been around for a very, very long time. But our specific interest in rural and small town places begins with the processes associated with the significant industrial resource-development activities and investments that emerged after the Second World War. This public policy and capital investment undertaking set in place resource tenures, physical and industrial infrastructure, resource-dependent towns, and workforces. It reorganized local, regional, and to a degree state, economies and societies to meet the needs of the industrial resource-development model. On this landscape, there have been since 1980 new sets of pressures.

To start, the demands for consumer goods that the post-war baby-boom generation within developed economies created is now being replaced by a demand for consumer goods by the growing 'middle-class' in China and India. While this will continue to support the demand for natural resource commodities, the price competition on those commodities from low-cost producer regions has not only complicated the market place for traditional suppliers, but it has also supported capital flight as multi-national firms have relocated production from developed economies to low-cost regions, especially those in the global south. While post-war resource development sought to address economic development needs, it was also a way to absorb surplus labour. Today, not only are so many developed economies facing lower birth rates and a declining labour pool, but the skills needs and training for resource-industry jobs are also much different today than in the past. Matching continuing opportunities for resource development with the cost and availability of labour will be an important issue in those developed economies retooling their resource and small town regions. For resource-dependent rural and small town places, these matters add to the question of who will bear the cost, who will reap the benefits, and how can we re-imagine natural resource assets to achieve a better balance of the two for these places and regions.

Two trajectories that, while present for a much longer period, have risen in importance since the start of economic restructuring in the early 1980s, involve environmentalism and increasing awareness of Aboriginal / Indigenous rights and title. In the case of environmentalism, new resource-development projects in developed economies often encounter significant opposition over concerns about the short- and long-term environmental impacts of the activity. Established resource-development projects are also adjusting to the changing environmental monitoring processes and regulatory regimes brought about by similar pressures. Both established and new resource development are also part of emerging and larger debates and understandings of the cumulative community, environmental, and health impacts of natural resource development (Gillingham et al., 2016). But the rise of environmental consciousness also seems to have been geographically differentiated as mass mobilization of protests and boycotts are now common in developed economies, while in the less-developed regions around the world, the consequences of poor industrial practices or even the disposal of e-waste, to use a simple example, only raise occasional interest and protest. This geographic differentiation of environmental awareness, impacts, and protests exacerbates processes such as capital flight.

Wider awareness of Aboriginal / Indigenous rights and title is also transforming many resource-dependent rural and small town regions. Treaty, royalty, and revenue-sharing agreements, agreements on protected areas, resolution of matters such as ownership, attention to greater economic participation, and recognition of cultural and traditional practices are increasingly the normal working context in developed economies. As with opportunities for re-imagining and re-bundling assets, the opportunities these new relationships and understandings create are potentially significant for those regions. However, also like earlier limits to the role of the state, Aboriginal / Indigenous participation needs investment in many jurisdictions to overcome the negative legacies of colonialism.

Within the constraints of past decisions and investments, and recognizing the opportunities and the challenges of contemporary trajectories within our societies, it is very important that states and communities do not continue to commit investments aimed at replicating the immediate post-war industrial resource-development model. While resource extraction and export will continue to be an important part of developed economies for years to come, there are so many additional up- and down-stream opportunities, and so many chances to break free from path dependence, that new investments, programs, and policies must be aimed at supporting increasingly flexible infrastructure, smarter services, and greater community and individual capacity building. The creation of new competitive advantage will require a nimbleness of response which older governance and approval processes just could not cope with. For our rural and small town places, resilience and nimbleness must become the new watchwords.

Institutions

Following our introduction to the capitalist economy, and to the context of an increasingly globalized capitalist economy, we next turned our attention to a set of key institutional actors through which the implications of economic restructuring within natural resource industries flowed. As change accelerated after 1980, each of these institutional actors was transformed. Some responded quickly, while others have only adjusted incrementally.

The implications for large resource companies have included adjustments to the cost-price squeeze of low-cost producer regions entering the global marketplace. For the most part, economic imperatives have forced such private-sector firms to react relatively more quickly and deliberately than some of the other institutional actors profiled. While some firms have tried to cling to older product lines within older producer regions, most have adopted more flexible and innovative approaches with respect to both product lines and investment regions. The mobility which globalization and new information technologies has introduced have been harnessed by firms to renew, or to create new, economic strength. Looking forward, resource development within our subject rural and small town areas will continue to involve large industrial capital. This is because the scope of investments needed, together with connections in the marketplace, remain of importance. What will change, however, is the partnership arrangements with local communities around the sharing of, not only risk, but now also the benefits of any developments deemed to be in accord with their mutual interest. How these play out in different places and regions will be highly dependent upon those places and regions, and their aspirations and goals.

The story was somewhat the same for the small business sector within our rural and small town resource-dependent places. As noted, the economic structure of firms within these places was a unique mix of the very large and the very small. Long dependent upon a local industry for preferential purchasing and servicing arrangements, and on a 'captured' local workforce who would act as purchasers of consumer goods, small businesses have found both of these historically foundational pieces transformed. Resource industries now rarely give preferential treatment to local suppliers, looking instead to price and responsiveness factors when allocating contracts. The rise of long-distance labour commuting and of internet shopping have similarly upset the purchasing patterns of workers and their families. While not all have been successful by any means, there are many examples of innovative small business entrepreneurs changing their business model or their focus to take advantage of these more general transformations. The rural and small town place of the future will continue to see innovative small businesses linked into local and regional industry, and they will continue to see small businesses supplying the kinds of services and products that cannot be delivered so easily via an internet shopping company's aerial drones. The flexibility to respond to emergent opportunities will be key to a healthy small-business sector.

As noted for both major industry and small business alike, changes in the response to economic restructuring have had important implications for labour. Just as economic sectors have had to respond with flexibility, so too has labour. At times, this may be via changes to labour contracts and shop-floor structures. At other times, economic collapse or opportunity has required retooling, retraining, relocation, or the adoption of a long-distance labour commuting lifestyle. All of these types of changes have had important implications for workers, their families, and their communities. To the 1950s model of needing a good work ethic, workers must now add sets of specific skills, qualifications, and certifications. As economic sectors adapt, their workforces must even more become 'learning workforces'.

Job losses in the resource sector of developed economies have been a pervasive theme. Job growth around specific resource booms has also been a pervasive theme. It is the implications around sudden job losses, however, that is especially upsetting and can have catastrophic impacts on families and workers alike. While support services are part of the immediate response, the longer-term lesson for the future is that attention to ongoing training and capacity building will be increasingly important. How families react to such new realities of ongoing training, accelerated cycles of economic booms and busts, and the rise of long-distance labour commuting will also be a challenge, but will also be the norm moving forward.

While changes to industry, small business, and labour have been an important part of the adjustments to economic restructuring within the global resource sector, it is at the level of senior governments that the challenge of these adjustments has been the most problematic. As described earlier in the book, the success of the post-war industrial resource-development model was based on a clear recognition by senior governments of critical new directions and opportunities within the global economy, the identification of how the unique assets and capacities of their own jurisdictions could be adapted to fit with those new directions, and then the implementation of a long-term and holistic government policy framework to put in place both the economic and community development building blocks to realize those emergent opportunities. The successes of that policy framework lead, however, to the rise of two limiting factors. The first is that senior governments soon became 'addicted' to the significant wealth being generated from natural resource development. Too often, most of those senior government jurisdictions spent every penny of that new wealth as it came in, rather than developing a long-term trust or legacy fund. The second is that in the policy transformation from Keynesian to Neoliberal approaches, states lost much of their capacity to track critical shifts in their own economic foundations. The result of these limiting factors is that as the pressures of global economic restructuring impacted states, the state very often sought to shore up, or even expand, their faltering natural resource economies. The creative reimagining of their economic circumstances that industry, business, and labour under took was less apparent across the senior governments of most developed economy states.

Looking forward, senior governments must transform their policy frameworks if they are to be more successful in revitalizing and renewing their rural and small town resource-dependent regions. To start, there needs to be recognition that holistic and cross-government policy directions are needed instead of the current 'silver bullet' efforts. Attention to making investments in places so as to support their unique opportunities in regional, national, and global economies means equipping them with 21st-century infrastructure and services. It also means investing in capacity building and supporting greater collaborative governance. As noted earlier, natural resource production will continue as an economic sector (though a transformed economic sector), but the key will be to add diversification and resilience into local and regional economies. Place-based policy approaches will need to replace the 'one-size-fits-all' approach that is no longer suited to the complexity of rural landscapes. Investing to build competitive advantage in these landscapes will now mean attention to investing in people, communities, and economies collectively.

Compared to senior government, communities and their elected local governments are very much on the front lines when it comes to experiencing the changes and transformations of the global resources economy. As described earlier, rural and small town places along the resource-development frontier were often built new or expanded from existing small settlements as part of the post-war industrial resource model. This legacy has created significant challenges when these places face the realities of post-1980 economic restructuring. Among these challenges are that such places are generally small (thus limiting their internal capacity), generally located far from decision-making centres (thus making it difficult to get the attention of senior government policy and decision makers), will not make the final decision on key policy and economic investments (as these rest with senior governments and industry), and are equipped with a housing, service, and infrastructure base that was created in the post-war era and designed to house its relatively young work-force (and their baby-boom families). Today, the community and economic circumstances of these communities has changed along with the social, eco-nomic, demographic, policy, cultural, and related shifts more generally occurring within each of their national contexts. The stable economic founda-tions of these communities is increasingly less stable, workforces are older, infrastructure and service issues are more challenging, there are issues with recruiting younger workers and new economic activities, and so on. In some communities, there also the stresses that come when an economic boom strains community capacity.

Looking forward, rural and small town places will need to be attentive to their place in the global economy. They will need to be realistic about the status of their infrastructure, housing, and services base. From here, commu-nities will then need to maintain those dialogues that will help them to create new competitive advantage, and then the momentum to implement the changes necessary to realize that competitive advantage. As noted earlier, communities will need to engage in dialogue which re-thinks and re-imagines their assets, works to re-bundle those in creative and imaginative ways so as to create new economic and community development opportunities, and then ensure that these new opportunities fit with community goals and aspirations. Communities will also need to find ways to scale up with their partner communities to form more effective alliances in their relationships, and in their negotiations, with the state and with economic interests. Speaking with a collective voice is the only way small places will impact policy and development decision making.

As places cope and adjust to the new rural economy, we focused our attention on two groups in particular. The first involved in those civil society organizations comprised of voluntary-sector groups. Long a vital part of rural and small town resource-dependent communities, voluntary groups support many of the quality-of-life activities and services that are so important in the small places. As change has worked its way through these communities, these groups have been squeezed by relative declines in funding supports from industry and the state and by a relative increase in the level of demand for their services. Some of these increasing demands are the result of economic restructuring pressures, while others are a policy outcome of government retrenchment under the guise of Neoliberalism. As communities continue to experience economic and community change, voluntary groups will have larger and more important roles to play. In the coming decades, demographic aging will bring new demands, but also new pools of volunteers. To be successful, voluntary groups need to adapt not just their mandates as needs / demands shift, but also their ways of working and their ways of communicating. They will need to not only take advantage of the potential of social media and other ways of organizing for action, but also to stay relevant to new generations of community members who might themselves wish to give back to their communities. Local government and senior government can play roles in supporting the rural and small town voluntary sector. Capacity building, board training, sensible support programming, and less burdensome reporting on grants are only a few of the ways by which governments can help such groups be at their best when local communities need them the most.

Social geographies

As we constructed this political economy of change in resource-dependent rural and small town places and regions, we have done so from the perspective of social geographers. While numbers of social geographic issues have been explored throughout the book, in this section we focus on a couple of specific topics.

Colonialism

The post-Second World War expansion of the resource frontier continued, in many developed countries, both a physical and a metaphorical expansion of colonialism. The expansion of the industrial resource-development model into such settings often occurred without care and attention to Aboriginal / Indigenous populations or to their rights and title with respect to the land and its wealth. As a consequence, there was a continuation of the significant negative impacts that colonialism brings to traditional cultures, land-use practices, and life-styles. In too many places, among the most challenging of these continuing impacts has been de-territorialization and ongoing poverty in regions that have yielded great wealth for both industry and states.

Most states have now engaged in efforts of reconciliation relative to past colonialist practices. In some of these states, there has been significant progress made in terms of the legal rights and roles of Aboriginal / Indigenous peoples with respect to discussions and decisions regarding resource use and develop-ment. However, even where such has occurred, the processes too often still follow colonialist practices as external governments impose legal and other forms of participation and decision-making frameworks. It is very rare to find states and industries adopting de-colonialist practices and acceding to follow, and work within, traditional Aboriginal / Indigenous frameworks and world-views. Industries, local governments, and community service agencies are only now starting to adopt cultural awareness training and Aboriginal / Indigenous protocols for their workforces.

While work is still needed on legal frameworks and on the practical appli-cation of what recognition of Aboriginal / Indigenous rights and title means, work is also needed on resolving and addressing the continuing and long-term negative impacts of colonialism. To begin to address these impacts, long-term attention is needed to supporting processes of healing and recovery. Support is also needed for cultural and traditional practices. Addressing many of the lagging quality-of-life issues within Aboriginal / Indigenous communities is also critically needed. Such issues include: healthy and appropriate housing; quality water, storage, transportation, and communications infrastructure; and health, educational, social service, and recreational services and facilities. More important still is the need to build individual and community capacity in ways that supports future engagement in both traditional and capitalist economies and societies so that choice can be exercised. The negative impacts of colonialism have been around for a long time, and it will take a long time to address them, but it is in the interest of all that they be addressed.

Gender

The topic of gender has long been of interest to social geographers. At its simplest, 'gender' does not refer to biological differences between men and women. Rather it refers to the interactions of socially constructed differences between men and women. It includes a socially organized and regulated world of norms, attitudes, behaviours, and opportunities. It also includes the various processes of socialization and acculturation that reinforce, and act to change, gender relations in places over time. Recent debates about the personal and wider social construction of gender and sexual identities, including transgender identities, highlight how the topic of gender and gender relations is a contested one, complete with ongoing change, debate, and conflict.

Some of the foundational bases for contention and conflict around gender norms and relations has to do with general processes of change to the status quo that create uncertainty, but more to the point, they are also often part of the reproduction of power relations. As changes in gender norms and expectations flow into societies and economies, changes in the power structures within those societies and economies create focal points for debate. Protection of political or economic power, disputes over eligibility for certain types of employment or rights of access, etc., are typical of these focal points for contention and conflict.

Against this conceptual backdrop, gender relations in resource-dependent places and regions have some relatively common traits. Through the post-war industrial resource-development era, employment opportunities, especially within the resource sector, were dominated by men. Men tended to be employed more often in high-paying, full-time, generally union-organized, jobs within the local mills, mines, or harvesting / extracting activities. Women, on the other hand, generally reported lower levels of labour-force participation. Some of this was due to the demographic structure of the baby-boom families of the era, but there was reinforcement of gendered norms via a host of social and economic processes. When women did work, they tended to be employed more often in low-wage, part-time, service-sector jobs without union protection. As a result, women in these types of communities more often than men found themselves in relatively precarious work circumstances. Professional opportunities for women were also found in the educational, health care, and local-government sectors. One of the many curiosities of the post-war employment period is that male employment typically did not require any special educational attainment and many young men secured high-paying resource-sector jobs even after dropping out of high school. In contrast, better jobs for women often required specific post-secondary training and trends found that women in these places were on the whole better educated than men.

The masculinization of work opportunities (read 'economic power') permeated these types of communities and shaped both household dependencies and community social relations. The economic divisions it supported also created significant local division and dependence. Low-income households struggled where the price of goods, services, and housing was generally geared towards the high-wage resource sector. Within households, the economic power generally rested with men. Differences in work opportunities also translated into differences in community participation. In this case, women often played key roles in the voluntary sector and related forms of community leadership.

While the stereotypical image of male workforce and female caregivers at home has always been problematic and has certainly been breaking down in recent decades, Reed (2008) highlights how the older stereotypes have never fully captured the complexity of women's participation in the resource-sector labour force. On the other hand, Ekers (2013, p. 877) notes that, even today, gender and sexuality remain central to social constructions of work in the contract tree planting sector where he found that management and workers

> create a charged heterosexualized working culture in an effort to bolster production levels ... [which results] ... in the gendering of work as men are framed as the protagonists of production while women are subordinated to a secondary position within the industry.

Today, and looking forward, economic restructuring also sets in place numbers of implications for women's participation in the labour force of resource-dependent regions. To start, outright job reductions and the increasing shift from full-time towards part-time work in the resource sector, whether through the increased use of flexible labour (part-time or on-call) or contract labour, is rolling back opportunities for women to enter the world of 'high-paying full-time' jobs held by their older male counterparts. That new jobs being created within these sectors are increasingly part-time also limits opportunities for women to enter the 'high-paying full-time' jobs held by their older male counterparts. Given that women historically formed the largest segment of part-time workers, these impacts are important. In some cases, however, there is considerable evidence of increasing opportunities for women to participate in high-wage resource-sector jobs. As more of these jobs require less physical labour and more technical skills, a more level playing field is beginning to develop. Haul truck driving, heavy equipment operations, control room management, and work as registered professional foresters are simply a few of the areas where this level playing field is being experienced.

The implications have been, and will be, equally significant for men. Where jobs acted as surrogates not only for place in the economy and in the community, but also for place within the family and within relationships, the past three decades have created a seismic upheaval in constructions of masculinity within resource-dependent places. As dual-income households become the norm, industrial restructuring has meant that increasingly both spouses are engaging in long-distance commuting practices when local industries idle their operations. The impacts on identity and employment have consequences that, as yet, have not really been addressed within the economic and community development supports provided around the restructuring of resource industries and a resource-dependent rural and small town places and regions.

Aging

Another social geography topic concerns that of population aging. In most developed economies, the post-war 'baby boom' established a very specific demographic 'wave' within the population that has 'moved up' the local, regional, or national population pyramids as time passed. Due to its sheer size, the baby-boom cohort created sets of social and economic impacts within society. In terms of the economy, for example, their impact on consumer goods of different types was not only significant, but it also shifted and changed as the population aged. This population growth also established much larger markets for any goods and services that were produced – a perfect companion of the expanded industrial capacities and capabilities that resulted from the Second World War.

There were also social impacts as well. As children, the baby-boom generation was part of the nuclear family, suburban neighbourhood, 'stay-at-home mom' phenomenon that shaped places and societies. As they aged, the sheer size of their cohort demanded attention to their stages of life – the rebellious teens, the experimental college years, climbing the corporate ladder, family formation, etc. Today, with the baby-boom generation entering retirement, attention now shifts to redefining the meanings of aging and retirement, as well as the expectations held around life and lifestyle among older people.

While the baby-boom generation is a significant 'macro' demographic process within developed economies, there is another process that complicates the future demography of resource-dependent rural and small town places and regions. The post-war economic boom and expansion of the industrial resource model into more rural and remote regions in the 1950s and 1960s greatly increased the numbers of new job opportunities in these areas. These job opportunities attracted young workers who, when they found good-paying jobs in stable and secure industries, began to establish young families. As a result, resource-dependent regions came to have communities dominated by very young populations. In the case of newly built communities, along the resource frontier, there were often very few older residents, let alone any retired individuals.

Over time, the workforce in these new resource industries began to age with the industries. As we have seen, increasing competition from other regions slowed the development of new industrial facilities such that the creation of large numbers of job opportunities that was a characteristic of such places began to dry up. As the 1970s transitioned into the global economic recession of the early 1980s, and economic restructuring began to impact established resource industries, the earlier described processes of job losses and economic contraction resulted. The long-term impacts of these processes have reinforced workforce aging-in-place and supported significant out-migration among younger people, due to the lack of new local job opportunities. Hanlon and Halseth (2005) described this process of 'resource-frontier aging' where the general demographic of baby-boom population aging is exacerbated by the processes of industrial change in the resource sector.

Interest in the social geography of aging within resource-dependent places has been expanding (Skinner and Hanlon, 2016). A key challenge identified in this growing literature is the fact that most community development investment in resource-dependent communities occurred early in the post-war industrial cycle, when the demographic make-up of these communities was focused on young families and young children. Infrastructure, services, housing, etc., were all geared to the needs of young people. As Hanlon et al. (2016, p. 11) note, there will be significant attention and effort needed to "transform the physical infrastructure and social systems of resource-dependent places in order to accommodate a growing population of older residents".

As noted earlier, not only are older residents a matter of social concern, but they are also a matter of economic concern as they contribute an increasingly important part of the basic sector economic diversity of many communities. Attention to the issues of transforming infrastructure and services so as to be more 'age-friendly', is an important part of the formula to holding local retirees in resource-dependent rural and small town places for some, if not all, of their older years. There is also the question of whether such places can be attractive to older migrants and grow this part of the social and economic base. As people age, there are different motivations at different times for their mobility. School, work, family formation, the search for work, the search for better incomes, the search for a better quality of life, all dominate at different points in the life-span of households in developed economies. At retirement, some different motivations begin to appear. Key among these are the locations of family and friends – with the locations of grandchildren being especially important for older households with the financial freedom to relocate. This speaks to the need for resource-dependent rural and small town places to be able to not only hold older cohorts, but also to create employment opportunities for younger households – and especially in light of the goals and desires of the next generation workers described earlier in the book. Rural and small town places have shown themselves to be very competitive for the quality-of-life factors that many next generation workers are looking for. If communities themselves can invest in the services, amenities, and housing to support people of all ages and in all stages of life, the social and the economic will have come together successfully.

Another interesting older cohort includes those raised in, or who raised families in, resource-dependent rural and small town places and regions. Upon retirement, can they be attracted back to these places. Jauhiainen (2009) found that in Finland, not only could return migration trigger additional local development, but that these places have many of the attractive factors that older households are seeking. They include "clean nature, peaceful environment, security, detached houses with beautiful setting, lower housing costs and landscapes of home" (Jauhiainen, 2009, p. 25).

As part of the discussion about older residents in small places, and the need to provide 'different' services than had traditionally been the case, there is concern about the burden of care demands that maybe created. These are important matters and the general research on population aging highlights two aspects of this topic (Wiles and Jayasinha, 2013). At the one end, there are the service needs that increase as people age. On the other hand, older people are active contributors to the social supports required for healthy population aging and trends suggest that the most intensive need for medical and care services is being pushed into a smaller number of years very close to the end of life. Skinner and Hanlon (2016) pursue this debate in the specific context of resource-dependent rural and small town places. They highlight the realities of demographic change within the economic histories of such communities and the contributions of older residents to service provision through their voluntary capacity. They argue for a more nuanced, place-based, and embedded understanding of population aging and its social and economic implications in these types of places.

Second home / amenity migration

Connected in part to processes of population aging are those processes related to retirement migration. A well-established retirement migration pattern in most developed economies is for the household to relocate to a high-value amenity setting – perhaps even a site where they own vacation property or have enjoyed vacationing in the past. Such migrations are, of course, circumscribed and bounded by relative household wealth.

In rural and small town places with amenity-rich environments that include shorelines (oceans or lakes), rivers, mountains, etc., the phenomenon of second-home recreational property development is certainly not new (Hall and Müller, 2004). As economic restructuring has increasingly impacted resource-dependent regions, retirement migration, or amenity migration (a different variation on this phenomenon that is additionally described in the various migration literatures), has often been identified as a potential way to diversify local and regional economies. This option is sometimes mobilized with other more general tourism-development strategies, and sometimes it is mobilized completely separately.

Such amenity or retirement migration to rural and small town areas with a mix of permanent and seasonal residents does create additional service needs as roads and water / sewer services are put under additional strain and may need to be upgraded or expanded. The introduction of large numbers of newcomers also brings changes in the local social dynamic and political organization, and to the visions for the future of the community that different groups may desire (Halseth, 1998). As noted by Kondo et al. (2012, p. 174), second-home owners or amenity migrants actively "seek to protect their investments by supporting regulations which support their vision of the rural idyll." This is very different from historical models, where the 'rural' and the 'recreational' lived quite separate from one another in many amenity-rich regions, despite the fact that they essentially occupied very similar space.

Creative class

In looking forward, it is useful to bring together a number of threads from our earlier discussions. As economic, social, political, cultural, and environmental change unfolds, resource-dependent rural and small town places are retooling by using their place-based assets to create new competitive advantage opportunities within the global economy. As population aging unfolds, these same places are adjusting their service, housing, and infrastructure components to not only hold older residents, but also to be attractive to older residents as part of social, economic, and cultural development strategies. This attention, as we noted earlier, must also be mindful of the need to attract younger households and individuals to take up local employment opportunities as the area's workforce begins to retire in large numbers. But we have also identified another economic development opportunity for non-retired households – those who wish to live in rural and small town places but who have the capacity and flexibility to work globally, either through telecommuting or routine long-distance labour commuting.

Part of this phenomenon has to do with households who have transferable skills and who may wish to remain in, or relocate to, rural and small-town places. Another part has to do with the amenity migrants identified earlier. But there is still further opportunity as one of the groups of workers and economic activity that has received a lot of press involves the 'creative class'. Commensurate with the transitioning of the economy from goods producing to one focused on service, information, and management activities, the mental labour of talented people is replacing aspects of the physical labour that had for so long defined work opportunities in resource-dependent rural and small town places. While much of the literature on creative economies and the creative class has focused on large urban centres (Florida, 2002), opportunities in rural and small town places have also been noted (Bruce, 2010).

Kourtit et al. (2011) highlight the convergence of creative economic opportunities with the concept of clusters. This fits with our earlier arguments around re-imagining and re-bundling place-based assets, including quality of life, community infrastructure, educational institutions, etc. in order to support creative economic opportunities in localities where people also wish the attractive features of rural and small town places. As fits the general argument, "the Creative Class seek out tolerant, diverse, and open communities, rich in the kind of amenities that allow them precariously to maintain a work-life balance, together with experiential intensity, in the context of those demanding work schedules" (Peck, 2005, p. 745). Identified for urban places, and in moving forward with opportunities, Kourtit et al. (2011) echo our call for place-based policy supports, investments in local services and amenities, investments in labour-force skills, and attention to the synergies of human and social capital / social cohesion as parts of the search by rural and small town places and regions for new competitive opportunities to diversify the local community and economy, and to build resilient capacity.

Services

Pulling together many of the topics discussed in the social geography of change in resource-dependent rural and small town places and regions is that of services and service infrastructure. As noted earlier, one of the responses of senior governments to the double impacts of industrial restructuring and Neoliberal policy transition has been to reduce expenses and rural service delivery, reconsider monies for critical infrastructure as expenses rather than long-term investments, and to download responsibility for services and infrastructure onto local communities. As noted in Markey et al. (2012), the approach has been degrading the service and infrastructure base for rural and small town places, at precisely the time when those places need increasing investment in order to support the new local / regional entrepreneurial approaches to economic development being advocated for by those very same senior governments. Pressures of profitability on the private sector have similarly led to the closure or regionalization of rural and small town services.

While numerous studies from different jurisdictions have highlighted the challenges confronting rural and small town service delivery and infrastructure, the need for attention to service and infrastructure investment is even more widely recognized. A report by the Business Council of British Columbia (2014, p. i) identified that "additional infrastructure investments are necessary to support residents' quality of life and improved BC's competitive position", and that this is

> an especially important consideration for a small, trade-oriented jurisdiction such as BC [given that] the world has become more competitive. To succeed in the global marketplace, BC must have top quality infrastructure that supports the development of human capital and facilitates business operations and trade.

Industries are also recognizing the importance of services and supports. As Neoliberal public policies have reduced service-sector supports in rural and small town resource-dependent regions, industries are pursuing service investments of their own. For example, with the growth of mobile workforces, industries are providing / contracting their own health care services in an effort to address gaps in mental health and wellness supports for these work-forces. While many work camps have paramedics and nurses, there are also work camps and industries that are using the FIFO model to engage doctors and related health care professionals; in some cases, even providing rooms for locum doctors visiting nearby communities (Ryser et al., 2015).

Two issues continue to be raised in the rural service literature: the need for 'generalist' services and the need for 'smarter' services. In the generalist ser-vice literature, there are also arguments being made for breaking service monopolies. This has been especially noticeable in the health care profession and discussions focused on the potential of expanded scopes of practice for nurses, paramedics, and others in under-serviced rural areas (Stirling et al., 2007).

There are many different terms used for smart services and infrastructure, including smart services, smart partnerships, one-stop shops, coordination initiatives, multi-service agencies, multi-purpose centres, co-location, single-window services, single gateways, multi-purpose service programs, and the like. Whatever the label, when it comes to the smarter and more efficient delivery of services, suggestions have been many. They include one that has significant potential for rural and small town resource-dependent places – the consolidation of services into multi-use facilities. This not only reduces general operating costs for the services themselves, but it creates a convenient one-stop shop for area residents. Discussion of 'smarter' service provision has also included a focus on the use of new technologies to deliver services in rural and remote areas or in small town settings where the smaller number of service users precludes local delivery of services. One set of models make use of older technologies such as telephone lines and connections to service call centres, while others focus on the use of internet-based service provision (Cassidy, 2011).

Strengthening the transition for those communities that were historically resource-dependent and that are now working to compete in the new rural economy will be shaped in part by the availability of a supportive service sector that can address the continuum of needs for the next generation workforce, the existing workforce, and to retain the wealth of the retiring workforce that has invested much of their expertise, experience, and resources in these resource-dependent regions. A strong services base will help with transition processes themselves and will also add capacity for the dialogue needed to support the reimaging of local assets and the re-bundling of those assets to create competitive advantage. The success of the industrial resource-development model in the decades immediately after the Second World War was based in good part on the availability of high-quality services for the businesses, industries, and workforces who relocated to the resource frontier. As these communities, economies, and regions transition, a 21st-century suite of services appropriate for the community and the economy will be required.

Investments in that earlier era paid dividends for generations; what is needed now is a renewed commitment to investments in a services base that will pay dividends to future generations.

Closing

It's an old world, and it's a new world. As resource-dependent rural and small town places and regions within developed economies look forward to the future, they confront both opportunity and challenge. They confront a world still hungry for natural resources, even while substitution and competition from low-cost regions may attract market investment. They confront a world where both capital and labour have surprising levels of mobility supported by the new policy and trade regimes of globalization. They confront a world where the state, which had previously supported those places and regions through investment and infrastructure, is now limited by its own Neoliberal public policy approach, such that communities, regions, and their supporting social and economic infrastructure are not being retooled for competitiveness in the 21st-century.

The political economy of resource-dependent rural and small town places and regions has been shaped by powerful social, economic, and political forces. In the 1950s to 1970s period of industrial resource expansion within developed economies, these forces set in place both relationships and landscapes. As change began to impact all of these forces, albeit at different rates and with different intensities, issues of conflict and contention became the norm as part of the contest over who would exercise power and influence in the reshaping of those established relationships and landscapes. In the reframing of a new political economy of rural and small town places, new social and economic players are now involved, there are new societal attitudes with respect to the environment and Aboriginal / Indigenous rights and title, new governance arrangements are increasingly common, and there is a wider base of knowledge about how such communities and economies can adjust to pressures, seize opportunities, and create new futures. The political economy of formerly resource-dependent places will involve more interests, be more dynamic and flexible, and will be one where change is acknowledged not as an exception but as the norm.

Bibliography

Alaska Permanent Fund Corporation. 2016. *Alaska Permanent Fund*. Available online at: http://www.apfc.org/homeobjects/tabPermFund.cfm. Accessed August 31, 2016.

Ali, M. 1986. "The coal war: Women's struggle during the miners' strike". In R. Ridd and H. Callaway (eds.), *Caught up in conflict* (pp. 84–105). London: Macmillan Education.

Anglo American Services. 2012. *SEAT toolbox: Socio-economic assessment toolbox, version 3*. London, UK: Anglo American Services.

Amin, A. 2009. "Institutionalism". In D. Gregory, R. Johnston, G. Pratt, M. Watts, and S. Whatmore (eds.), *The dictionary of human geography*, 5th edition (pp. 386–387). Oxford: Wiley-Blackwell.

Argent, N. 2013. "Reinterpreting core and periphery in Australia's mineral and energy resources boom: An Innisian perspective on the Pilbara", *Australian Geographer* 443: 323–340.

Argent, N. 2017a. "Labour/capital relations and sustainable development in the NSW Northern Tablelands". In G. Halseth (ed.), *Transformation of resource towns and peripheries: Political economy perspectives* (pp. 142–160). Oxford and New York: Routledge.

Argent, N. 2017b. "Trap or opportunity? Natural resource dependence, scale, and the evolution of new economies in the space/time of New South Wales' Northern Tablelands". In G. Halseth (ed.), *Transformation of resource towns and peripheries: Political economy perspectives* (pp. 18–50). London and New York: Routledge.

Armenakis, A. and Bedeian, A. 1999. "Organizational change: A review of theory and research in the 1990s", *Journal of Management* 253: 293–315.

Asquith, A. 2012. "The role, scope and scale of local government in New Zealand: Its prospective future", *Australian Journal of Public Administration* 71(1): 76–84.

Barca, F., McCann, P., and Rodríguez-Pose, A. 2012. "The case for regional development intervention: Place-based versus place-neutral approaches", *Journal of Regional Science* 52(1): 134–152.

Barnes, T. 1996. *Logics of dislocation: Models, metaphors, and meanings of economic space*. New York: Guilford Press.

Barnes, T. 2009. "Staples Theory". In D. Gregory, R. Johnston, G. Pratt, M. Watts, and S. Whatmore (eds.), *The dictionary of human geography*, 5th edition (pp. 721–722). Oxford: Blackwell Publishers.

Barr, C., McKeown, L., Davidman, K., McIver, D., and Lasby, D. 2004. *The rural charitable sector research initiative: A portrait of the non-profit and voluntary sector in rural Ontario.* Prepared for the Foundation for Rural Living. Toronto: Canadian Centre for Philanthropy.

Battilana, J. and Casciaro, T. 2012. "Change agents, networks, and institutions: A contingency theory of organizational change", *Academy of Management Journal* 552: 381–398.

Baxter, D. and Ramlo, A. 2002. "Resource dependency: The spatial origins of British Columbia's economic base". Vancouver: The Urban Futures Institute.

Bayari, C. 2016. "Economic geography of the Australian mining industry", *Tijdschrift voor economische en sociale geografie.* doi:10.1111/tesg.12185.

Beckley, T. 1996. "Pluralism by default: Community power in a paper mill town", *Forest Science* 42(1): 35–45.

Blake, R. 2003. *Regional and rural development strategies in Canada: The search for solutions.* St. John's, Newfoundland: Royal Commission on Renewing and Strengthening Our Place in Canada.

Bock, B. 2016. "Rural marginalization and the role of social innovation: A turn towards nexogenous development and rural reconnection", *Sociologia Ruralis* 56(4): 552–573.

Bollman, R. 2007. *Factors driving Canada's rural economy, 1914–2006.* Ottawa: Statistics Canada.

Boschma, R. and Frenken, K. 2006. "Why is economic geography not an evolutionary science? Toward an evolutionary economic geography", *Journal of Economic Geography* 6(3): 272–302.

Boschma, R. and Martin, R. (eds.). 2010. *Handbook of evolutionary economic geography.* Cheltenham, UK: Edward Elgar.

Bowles, P. 2013. "'Globalizing' northern British Columbia: What's in a word?", *Globalizations* 10(2): 261–276.

Bowles, P. and Wilson, G. (eds.). 2016. *Resource communities in a globalizing region: development, agency, and contestation in northern British Columbia.* Vancouver and Toronto: UBC Press.

Bradbury, J. and St. Martin, I. 1983. "Winding down in a Quebec mining town: A case study of Schefferville", *The Canadian Geographer* 27(2): 128–144.

Bradford, N. 2005. *Place-based public policy: Towards a new urban and community agenda for Canada.* Ottawa: Canadian Policy Research Networks.

Brand, J. 2015. "The far-reaching impact of job loss and unemployment", *Annual Review of Sociology* 41: 359–375.

Bruce, D. 2010. "Nurturing the animation sector in a peripheral economic region: The case of Miramichi, New Brunswick". In G. Halseth, S. Markey, and D. Bruce (eds.), *The next rural economies: Constructing rural place in global economies* (pp. 128–141). Wallingford, Oxfordshire: CAB International.

Bryant, C. 1999. "Community change in context". In J. Pierce and A. Dale (eds.), *Communities, development, and sustainability across Canada* (pp. 69–89). Vancouver: UBC Press.

Bryant, C. 2010. "Co-constructing rural communities in the 21st century". In G. Halseth, S. Markey, and D. Bruce (eds.), *The next rural economies: Constructing rural place in global economies* (pp. 142–154). Wallingford, Oxfordshire: CAB International.

Business Council of British Columbia. 2014. *Building BC for the 21st century: A White paper on infrastructure policy and financing.* Vancouver: Business Council of British Columbia.

Careless, J. 1989. *Frontier and metropolis: Regions, cities, and identities in Canada before 1914*. Toronto: University of Toronto Press.

Carrington, K. and Pereira, M. 2011. "Assessing the social impacts of the resources boom on rural communities", *Rural Society* 21(1): 2–20.

Cassidy, L. 2011. "Online communities of practice to support collaborative mental health practice in rural areas", *Issues in Mental Health Nursing* 32(2): 98–107.

Cater, J. and Jones, T. 1989. *Social geography: An introduction to contemporary issues*. New York: Halsted Press.

Cawley, M. and Nguyen, G. 2008. "Service delivery through partnerships in sparsely populated areas: Evidence from France and Ireland", *Irish Geography* 41(1): 71–87.

CBC News. 2016. "BC LNG prosperity fund to get $100M contribution, but not from LNG". February 15. Available online at: http://www.cbc.ca/news/canada/brit ish-columbia/lng-prosperity-fund-1.3449611. Accessed February 17, 2016.

CEDRA. n.d. *Commission for the economic development of rural areas research report*. Ireland: CEDRA; Teagasc Agriculture and Food Development Authority; Department of the Environment, Community, and Local Government; the Western Development Commission; and the Department of Agriculture, Food and the Marine.

Cheng, A., Kruger, L., and Daniels, S. 2003. "Place as an integrating concept in natural resource politics: Propositions for a social science research agenda", *Society and Natural Resources* 16(2): 87–104.

Cheshire, L. 2010. "A corporate responsibility? The constitution of fly-in, fly-out mining companies as governance partners in remote, mine-affected localities", *Journal of Rural Studies* 26(1): 12–20.

Chevron Australia. 2016. *Gorgon Project*. Available online at: https://www.chevrona ustralia.com/our-businesses/gorgon. Accessed August 30, 2016.

Clapp, A., Hayter, R., Affolderbach, J., and Guzman, L. 2016. "Institutional thickening and innovation: reflections on the remapping of the Great Bear Rainforest", *Transactions of the Institute of British Geographers* 41(3): 244–257.

Cloke, P., Johnsen, S., and May, J. 2007. "The periphery of care: Emergency services for homeless people in rural areas", *Journal of Rural Studies* 23(4): 387–401.

Cochrane, P. 2006. "Exploring cultural capital and its importance in sustainable development", *Ecological Economics* 57(2): 318–330.

Cohn, T. 2000. *Global political economy: Theory and practice*. New York: Longman.

Community Development Institute. 2007. *A summary of trusts in British Columbia*. Prince George, BC: Community Development Institute, University of Northern British Columbia.

Connelly, S. and Nel, E. 2017a. "Community responses to restructuring". In G. Halseth (ed.), *Transformation of resource towns and peripheries: Political economy perspectives* (pp. 317–335). Oxford and New York: Routledge.

Connelly, S. and Nel, E. 2017b. "Employment and labour in New Zealand: Recent trends and reflections on developments in the West Coast and Southland regions". In G. Halseth (ed.), *Transformation of resource towns and peripheries: Political economy perspectives* (pp. 221–241). Oxford and New York: Routledge.

Connelly, S. and Nel, E. 2017c. "Restructuring of the New Zealand economy: Global-local links and evidence from the West Coast and Southland regions". In G. Halseth (ed.), *Transformation of resource towns and peripheries: Political economy perspectives* (pp. 112–135). Oxford and New York: Routledge.

Connors, P. 2010. "Transition towns and community capacity building". In S. Kenny and M. Clarke (eds.), *Challenging capacity building* (pp. 229–247). New York: Palgrave Macmillan.

Cooke, D., Hill, C., Baskett, P., and Irwin, R. (eds.). 2014. *Beyond the free market: Rebuilding a just society in New Zealand*. Auckland, New Zealand: Dunmore Publishing.

Corbridge, S. 1994. "Maximizing entropy? New geopolitical orders and the internationalization of business". In G. Demko and W. Wood (eds.), *Reordering the world: Geopolitical perspectives on the 21st century* (pp. 281–300). Boulder, CO: Westview Press.

Courtney, P., Hill, G., and Roberts, D. 2006. "The role of natural heritage in rural development: An analysis of economic linkages in Scotland", *Journal of Rural Studies* 22(4): 469–484.

Crowe, J. 2006. "Community economic development strategies in rural Washington: Toward a synthesis of natural and social capital", *Rural Sociology* 71(4): 573–596.

Cumbers, A., MacKinnon, D., and McMaster, R. 2003. "Institutions, power, and space: Assessing the limits to institutionalism in economic geography", *European Urban and Regional Studies* 10(4): 325–342.

Dabscheck, B. 1994. "The arbitration system since 1967". In B. Head and S. Bell (eds.), *State, economy, and public policy in Australia* (pp. 142–168). Richmond, Australia: Heinemann Australia.

Davenport, M. and Anderson, D. 2005. "Getting from sense of place to place-based management: An interpretive investigation of place meanings and perceptions of landscape change", *Society and Natural Resources* 18(7): 625–641.

Davis, G. and Tilton, J. 2005. "The resource curse", *Natural Resources Forum* 29(3): 233–242.

DeFilippis, J. and Saegert, S. 2012a. "Communities develop: The question is, how?" In J. DeFilippis and S. Saegert (eds.), *The Community development reader*, 2nd edition (pp. 1–7). New York: Routledge.

DeFilippis, J. and Saegert, S. 2012b. "Concluding thoughts". In J. DeFilippis and S. Saegert (eds.), *The community development reader*, 2nd edition (pp. 377–382). New York: Routledge.

Del Casino, V. 2009. *Social geography*. West Sussex, UK: Wiley-Blackwell.

Department of Internal Affairs. 2015. *Community development scheme*. Auckland, New Zealand: Department of Internet Affairs. Available online at: http://www.communityma tters.govt.nz/Funding-and-grants—Crown-Funds—Community-Development-Scheme. Accessed December 22, 2015.

Department of State Development. 2016. *LNG profile*. East Perth, WA: Department of State Development, Government of Western Australia. Available online at: http://www.dsd.wa.gov.au/about-the-state/major-resource-producer/lng-profile. Accessed August 30, 2016.

Derkzen, P., Franklin, A., and Bock, B. 2008. "Examining power struggles as a signifier of successful partnership working: A case study of partnership dynamics", *Journal of Rural Studies* 24(4): 458–466.

Dewees, S., Lobao, L., and Swanson, L. 2003. "Local economic development in an age of devolution: The question of rural localities", *Rural Sociology* 68(2): 182–206.

Donnermeyer, J., Plested, B., Edwards, R., Oetting, G., and Littlethunder, L. 1997. "Community readiness and prevention programs", *Journal of the Community Development Society* 28(1): 65–83.

Douglas, D. 2005. "The restructuring of local government in rural regions: A rural development perspective", *Journal of Rural Studies* 21(2): 231–246.

Doussard, M. and Schrock, G. 2015. "Uneven decline: Linking historical patterns and processes of industrial restructuring to future growth trajectories", *Cambridge Journal of Regions, Economy and Society* 8(2): 149–165.

Drache, D. 1991. "Harold Innis and Canadian capitalist development". In G. Laxer (ed.), *Perspectives on Canadian economic development: Class, staples, gender, and elites* (pp. 22–49). Don Mills: Oxford University Press Canada.

Drohan, M. 2012. *The 9 habits of highly effective resource economies: Lessons for Canada.* Toronto: Canadian International Council.

Dublin, T. 1979. *Women at work: The transformation of work and community in Lowell, Massachusetts, 1826–1860.* New York: Columbia University Press.

Dufty-Jones, R. and Wray, F. 2013. "Planning regional development in Australia: Questions of mobility and borders", *Australian Planner* 50(2): 109–116.

Dumais, S. 2002. "Cultural capital, gender, and school success: The role of habitus", *Sociology and Education* 75(1): 44–68.

Edenhoffer, K. and Hayter, R. 2013. "Organizational restructuring in British Columbia' forest industries 1980–2010: The survival of a dinosaur", *Applied Geography* 40: 222–231.

Education and Health Standing Committee. 2015. *The impact of FIFO work practices on mental health.* Report No. 5. Perth: Legislative Assembly, Parliament of Western Australia.

Egan, B. and Klausen, S. 1998. "Female in a forest town: The marginalization of women in Port Alberni's economy", *BC Studies* 118: 5–40.

Ekers, M. 2013. "'Pounding dirt all day': Labor, sexuality and gender in the British Columbia reforestation sector", *Gender, Place and Culture* 20(7): 876–895.

Ellem, B. 2006. "Scaling labour: Australian unions in the old economy: Community and collectivism in the Pilbara's mining towns", *The Journal of Industrial Relations* 45: 423–441.

Emery, M. and Flora, C. 2006. "Spiraling-up: Mapping community transformation with community capitals framework", *Community Development* 37(1): 19–35.

England, K. 1993. "Suburban pink collar ghettos: The spatial entrapment of women?", *Annals of the Association of American Geographers* 83(2): 225–242.

Essletzbichler, J. and Rigby, D. 2007. "Exploring evolutionary economic geographies", *Journal of Economic Geography* 7: 549–571.

Fagan, R. and Webber, M. 1999. *Global restructuring: The Australian experience*, 2nd edition. South Melbourne, Australia: Oxford University Press.

Fidler, C. and Hitch, M. 2007. "Impact benefit agreements: A contentious issue for Aboriginal and environmental justice", *Environments Journal* 35(2): 49–69.

Fitchen, J. 1991. *Endangered spaces, enduring places: Change, identity, and survival in rural America.* Boulder, CO: Westview Press.

Flint, C. 2009. "World-systems analysis". In D. Gregory, R. Johnston, G. Pratt, M. Watts, and S. Whatmore (eds.), *The dictionary of human geography*, 5th edition (pp. 813–814). Oxford: Wiley-Blackwell.

Florida, R. 2002. *The rise of the creative class: And how it's transforming work, leisure, community and everyday life.* New York: Basic.

Fløysand, A. and Jakobsen, S. 2016. "In the footprints of evolutionary economic geography", *Norsk Geografisk Tidsskrift-Norwegian Journal of Geography* 70(3): 137–139.

Fogarty, J. and Sagerer, S. 2016. "Exploration externalities and government subsidies: The return to government", *Resources Policy* 47: 78–86.

Forum of Labour Market Ministers. 2015. *Provincial-Territorial Apprentice Mobility Agreement.* Available online at: http://www.flmm-fmmt.ca/CMFiles/Signed%20Agreement-EN.PDF. Accessed November 10, 2016.

Foster, J. and Taylor, A. 2013. "In the shadows: Exploring the notion of community for temporary foreign workers in a boomtown", *Canadian Journal of Sociology* 38(2): 167–190.

Fowler, K. and Etchegary, H. 2008. "Economic crisis and social capital: The story of two rural fishing communities", *Journal of Occupational and Organizational Psychology* 81(2): 319–341.

Franks, D. 2012. *Social impact assessment of resource projects.* International Mining for Development Centre. Crawley, Western Australia: Australia University of Western Australia.

Freudenburg, W. 1992. "Addictive economies: Extractive industries and vulnerable localities in a changing world economy", *Rural Sociology* 57(3): 305–332.

Freudenburg, W. and Gramling, R. 1994. "Natural resources and rural poverty: A closer look", *Society and Natural Resources* 7(1): 5–22.

Garrod, B., Wornell, R., and Youell, R. 2006. "Re-conceptualising rural resources as countryside capital: The cause of rural tourism", *Journal of Rural Studies* 22(1): 117–128.

Gibson-Graham, J. 1996. *The end of capitalism as we knew it: A feminist critique of political economy.* Oxford: Blackwell Publishers.

Gidwani, V. 2009. "Marxist economics". In D. Gregory, R. Johnston, G. Pratt, M. Watts, and S. Whatmore (eds.), *The dictionary of human geography*, 5th edition (pp. 445–446). Oxford: Wiley-Blackwell.

Gill, A. 1990. "Women in isolated resource towns: An examination of gender differences in cognitive structures", *Geoforum* 21(3): 347–358.

Gillingham, M., Halseth, G., Johnson, C., and Parkes, M. (eds.). 2016. *The integration imperative: Cumulative environmental, community and health impacts of multiple natural resource developments in northern British Columbia.* New York: Springer.

Government of Iceland. 2002. *Memorandum of Understanding between the Government of Iceland and Landsvirkjun and Alcoa Inc.* Available online at: https://eng.atvinnuvegaraduneyti.is/media/Acrobat/MOU_Alcoa.PDF. Accessed November 10, 2016.

Government of Western Australia. 2010. *Browse LNG precinct strategic social impact assessment: volume 3: Strategic social impact management plan.* East Perth, WA: Department of State Development.

Graddy, E. and Morgan, D. 2006. "Community foundations, organizational strategy, and public policy", *Nonprofit and Voluntary Sector Quarterly* 35(4): 605–630.

Grafton, D., Troughton, M., and Rourke, J. 2004. "Rural community and health care interdependence: An historical, geographical study", *Canadian Journal of Rural Medicine* 9(3): 156–163.

Gregory, D. 1978. *Ideology, science, and human geography.* London: Hutchinson.

Hackett, R. 2013. "From government to governance? Forest certification and crisis displacement in Ontario, Canada", *Journal of Rural Studies* 30: 120–129.

Haley, B. 2011. "From staples trap to carbon trap: Canada's peculiar form of carbon lock-in", *Studies in Political Economy* 88(1): 97–132.

Hall, C. and Müller, D. (eds.). 2004. *Tourism, mobility, and second homes: Between elite landscape and common ground*. Clevedon, UK: Channel View Publications, Vol. 15.

Halonen, M., Vatanen, E., Tykkyläinen, M., and Kotilainen, J. 2017. "Industrial labour in a resource town in Finland: The case of Lieksa". In G. Halseth (ed.), *Transformation of resource towns and peripheries: Political economy perspectives* (pp. 195–220). Oxford and New York: Routledge.

Halseth, G. 1998. *Cottage country in transition: A social geography of change and contention in the rural-recreational countryside*. Montreal: McGill-Queen's University Press.

Halseth, G. 1999. "'We came for the work': Situating employment migration in B.C.'s small, resource-based, communities", *The Canadian Geographer* 43(4): 363–381.

Halseth, G. 2005 "Resource town transition: Debates after closure". In S. Essex, A. Gilg, R. Yarwood, J. Smithers, and R. Wilson (eds.), *Rural change and sustainability: Agriculture, the environment and communities* (pp. 326–342). Wallingford, Oxfordshire, UK: CAB International.

Halseth, G. 2016. "The changing nature of resource economies: A focus on the example of forestry". In M. Shucksmith and D. Brown (eds.), *Routledge international handbook of rural studies* (pp. 108–119). London: Routledge.

Halseth, G. (ed.). 2017. *Transformation of resource towns and peripheries: Political economy perspectives*. Oxford and New York: Routledge.

Halseth, G. and Ryser, L. 2007. "The deployment of partnerships by the voluntary sector to address service needs in rural and small town Canada", *Voluntas* 18(3): 241–265.

Halseth, G. and Sullivan, L. 2002. *Building community in an instant town: A social geography of Mackenzie and Tumbler Ridge, British Columbia*. Prince George, BC: UNBC Press.

Halseth, G., Sullivan, L., and Ryser, L. 2003. "Service provision as part of resource town transition planning: A case from Northern British Columbia". In D. Bruce and G. Lister (eds.), *Opportunities and actions in the new rural economy* (pp. 19–46). Sackville, New Brunswick: Rural and Small Town Programme.

Halseth, G., Markey, S., and Bruce, D. (eds.). 2010a. *The next rural economies: Constructing rural place in a global economy*. Wallingford, Oxfordshire, UK: CAB International.

Halseth, G., Markey, S., Reimer, B., and Manson, D. 2010b. "Space to place: Bridging the gap". In G. Halseth, S. Markey, and D. Bruce (eds.), *The next rural economies: Constructing rural place in a global economy* (pp. 1–16). Wallingford, Oxfordshire, UK: CAB International.

Halseth, G., Manson, D., Ryser, L., Markey, S., and Morris, M. 2015. "Constructing rural places in a globalized world: Place-based rural development seen from northern British Columbia, Canada". In T. Gjertsen and G. Halseth (eds.), *Sustainable development in the Circumpolar North – From Tana, Norway to Oktemtsy, Yakutia, Russia: The Gargia conferences for local and regional development 2004–14* (pp. 227–239). Prince George, BC: Publications Series of the UNBC Community Development Institute. Tromso, Norway: University Library at UiT The Arctic University of Norway, Septentrio Conference Series, Number 1, 2015.

Halseth, G., Gillingham, M., Johnson, C., and Parkes, M. 2016. "Cumulative effects and impacts: The need for a more inclusive, integrative, regional approach". In

M. Gillingham, G. Halseth, C. Johnson, and M. Parkes (eds.), *The integration imperative: Cumulative environmental, community and health impacts of multiple natural resource developments in northern British Columbia* (pp. 3–20). New York: Springer.

Halseth, G., Ryser, L., and Markey, S. 2017a. "Building for the future: Community responses to economic restructuring in Mackenzie, BC". In G. Halseth (ed.), *Transformation of resource towns and peripheries: Political economy perspectives* (pp. 268–295). Oxford and New York: Routledge.

Halseth, G., Ryser, L., and Markey, S. 2017b. "Contentious flexibility: Job losses in labour restructuring in Mackenzie, BC". In G. Halseth (ed.), *Transformation of resource towns and peripheries: Political economy perspectives* (pp. 161–194). Oxford and New York: Routledge.

Halseth, G., Ryser, L., and Markey, S. 2017c. "Localization and globalization: Industrial re-organization in Mackenzie, British Columbia". In Halseth, G. (ed.), *Transformation of resource towns and peripheries: Political economy perspectives* (pp. 51–84). Oxford and New York: Routledge.

Hanlon, N. and Halseth, G. 2005. "The greying of resource communities in northern British Columbia: Implications for health care delivery in already underserviced communities", *The Canadian Geographer* 49(1): 1–24.

Hanlon, N., Skinner, M., Joseph, A., Ryser, L., and Halseth, G. 2014. "Place integration through efforts to support healthy aging in resource frontier communities: The role of voluntary sector leadership", *Health and Place* 29: 132–139.

Hanlon, N., Skinner, M., Joseph, A., Ryser, L., and Halseth, G. 2016. "New frontiers of rural ageing: Resource hinterlands". In: M. Skinner and N. Hanlon (eds.), *Ageing resource communities: New frontiers of rural population change, community development, and voluntarism* (pp. 11–23). London: Routledge.

Hanoa, R., Baste, V., Kooij, A., Sommervold, L., and Moen, B. 2011. "No difference in self reported health among coalminers in two different shift schedules at Spitsbergen, Norway: A two years follow-up", *Industrial Health* 49(5): 652–657.

Harris, M., Cairns, B., and Hutchinson, R. 2004. "So many tiers, so many agendas, so many pots of money: The challenge of English regionalization for voluntary and community organizations", *Social Policy and Administration* 38(5): 525–540.

Harris, R. 2002. *Making native space: Colonialism, resistance, and reserves in British Columbia*. Vancouver: UBC Press.

Harvey, D. 1975. *Social justice and the city*. Baltimore, MD: Johns Hopkins University Press.

Harvey, D. 1990. *The condition of postmodernity: An enquiry into the origins of cultural change*. Oxford: Basil Blackwell.

Harvey, D. 2005. *A brief history of neoliberalism*. Oxford and New York: Oxford University Press.

Harvey, D. 2010. "Roepke lecture in economic geography: Crises, geographical disruptions and the uneven development of political responses", *Economic Geography* 87(1): 1–22.

Haslam McKenzie, F. 2013. "Delivering enduring benefits from a gas development: Governance and planning challenges in remote Western Australia", *Australian Geographer* 44(3): 341–358.

Haslam McKenzie, F. and Rowley, S. 2013. "Housing market failure in a booming economy", *Housing Studies* 28(3): 373–388.

Hay, C. 2007. *Why we hate politics*. Cambridge: Polity Press.

Hayden, D. 1995. *The power of place: Urban landscapes as public history.* Cambridge, MA: MIT Press.

Hayter, R. 1982. "Truncation, the international firm, and regional policy", *Area* 14(4): 277–282.

Hayter, R. 2000. *Flexible crossroads: The restructuring of British Columbia's forest economy.* Vancouver: UBC Press.

Hayter, R. 2004. "Economic geography as dissenting institutionalism: The embedded-ness, evolution and differentiation of regions", *Geografiska Annaler Series B* 86(2): 95–1115.

Hayter, R. 2008. "Environmental economic geography", *Geography Compass* 2(3): 831–850.

Hayter, R. and Barnes, T. 2012. "Neoliberalization and its geographic limits: Comparative reflections from forest peripheries in the global north", *Economic Geography* 88(2): 197–221.

Hayter, R. and Holmes, J. 2001. "The Canadian forest industry: The impacts of glo-balization and technological change". In M. Howlett (ed.), *Canadian forest policy: Adapting to change* (pp. 127–156). Toronto: University of Toronto Press.

Heisler, K. and Markey, S. 2013. "Scales of benefit: Political leverage in the negotia-tion of corporate social responsibility in mineral exploration and mining in rural British Columbia, Canada", *Society and Natural Resources* 26(4): 386–401.

Herod, A. 1998. "The Spatiality of Labor Unionism: A review essay". In A. Herod (ed.), *Organizing the landscape: Geographical perspectives on labor unionism* (pp. 1–36). Minneapolis: University of Minnesota Press.

Hide, R. 2011. *Smarter government, stronger communities: Towards better local govern-ance and public services.* Wellington, New Zealand: Minister of Local Government, Department of Internal Affairs. Available online at: https://www.dia.govt.nz/diaweb site.nsf/Files/SGSCCabEGI113/$file/SGSCCabEGI113.pdf. Accessed December 22, 2015.

Hilmarsson, T. 2003. *Energy and aluminum in Iceland.* Landsvirkjun – National Power Company. Presented at the Platts Aluminum Symposium, Phoenix Arizona.

Hinde, J. 1997. "Stout ladies and Amazons: Women in the British Columbia coal-mining community of Ladysmith, 1912–1914", *BC Studies* 114: 33–57.

Holmlund, B. and Storrie, D. 2002. "Temporary work in turbulent times: The Swedish experience", *The Economic Journal* 112(June): 245–269.

Horsley, J. 2013. "Conceptualising the state, governance and development in a semi-peripheral resource economy: The evolution of state agreements in Western Australia", *Australian Geographer* 44(3): 283–303.

House, J. 1999. *Against the tide: Battling for economic renewal in Newfoundland and Labrador.* Toronto: University of Toronto Press.

House of Representatives. 2013. *Cancer of the bush or salvation for our cities? Fly-in, fly-out and drive-in, drive-out workforce practices in regional Australia.* Canberra, Australia: Standing Committee on Regional Australia, Parliament of the Commonwealth of Australia.

Hreinsson, E. 2007. "Deregulation, environmental and planning policy in the Icelandic renewable energy system". In International Conference on Clean Electrical Power. May: 283–290.

Hudson, R. 2006. "The 'new' economic geography?" In S. Bagchi-Sen and H. Smith (eds.), *Economic geography: Past, present, and future* (pp. 47–55). New York: Routledge.

Hultberg, E., Glendinning, C., Allebeck, P., and Lönnroth, K. 2005. "Using pooled budgets to integrate health and welfare services: A comparison of experiments in England and Sweden", *Health and Social Care in the Community* 13(6): 531–541.

Hunter, J. 2016. "BC budget to include 'prosperity fund' without LNG income". *Globe and Mail*February 15. Available online at: http://www.theglobeandmail.com/news/british-columbia/bc-budget-to-include-prosperity-fund-without-lng-income/article28762904. Accessed February 17, 2016.

Ingamells, A. 2007. "Community development and community renewal: Tracing the workings of power", *Community Development Journal* 42(2): 237–250.

Innis, H. 1933. *Problems of staple production in Canada.* Toronto: Ryerson Press.

Innis, H. 1956 [1929]. "The teaching of economic history in Canada". In M. Innis (ed.), *Essays in Canadian economic history* (pp. 3–16). Toronto: University of Toronto Press.

International Business Times. 2014. "Why Samsung's move to build smartphones in Vietnam will help it better compete in China". Available online at: http://www.ibtimes.com/why-samsungs-move-build-smartphones-vietnam-will-help-it-better-compete-china-1724014. Accessed November 10, 2015.

Ioris, A. 2016. "The politico-ecological economy of neoliberal agribusiness: Displacement, financialisation and mystification", *Area* 48(1): 84–91.

Ivanova, G. 2014. "The mining industry in Queensland, Australia: Some regional development issues", *Resources Policy* 39: 101–114.

Jackson, K. 1985. *Crabgrass frontier: The suburbanization of the United States.* New York: Oxford University Press.

Jakobsen, S. and Lorentzen, T. 2015. "Between bonding and bridging: Regional differences in innovative collaboration in Norway", *Norsk Geografisk Tidsskrift-Norwegian Journal of Geography* 69(2): 80–89.

Jakobsen, S., Byrkjeland, M., Båtevik, F., Pettersen, I., Skogseid, I., and Yttredal, E. 2012. "Continuity and change in path-dependent regional policy development: The regional implementation of the Norwegian VRI programme", *Norsk Geografisk Tidsskrift-Norwegian Journal of Geography* 66(3): 133–143.

Jauhiainen, J. 2009. "Will the retiring baby boomers return to rural periphery?", *Journal of Rural Studies* 25(1): 25–34.

Jean, B. 2014. *Innovation and modernising the rural economy.* Organisation for Economic Co-operation and Development.

Jessop, B. 2013. "Dynamics of regionalism and globalism: A critical political economy perspective", *Ritsumeikan Social Science Review* 5: 3–24.

Johnsen, S., Cloke, P., and May, J. 2005. "Transitory spaces of care: Serving the homeless people on the street", *Health and Place* 11(4): 323–336.

Joshi, M., Bliss, J., Bailey, C., Teeter, L., and Ward, K. 2000. "Investing in industry, underinvesting in human capital: Forest-based rural development in Alabama", *Society and Natural Resources* 13(4): 291–319.

Keenan, G. 2015. "Tax shift: Companies dump burden of taxes on squeezed municipalities". *The Globe and Mail*July 24. Available online at: http://www.theglobeandmail.com/report-on-business/tax-shift-companies-dump-burden-of-taxes-on-squeezed-municipalities/article25670719. Accessed February 2, 2016.

Kelsey, T., Shields, M., Ladlee, J., and Ward, M. 2012. *Economic impacts of Marcellus Shale in Bradford County: Employment and income in 2010.* Williamsport, PA: Marcellus Shale Education and Training Center, Pennsylvania College of Technology and Penn State Extension.

Kettner, P., Daley, J., and Weaver, A. 1985. *Initiating change in organizations and communities: A macro practice model.* Monterey, CA: Brooks / Cole Publishing Company.

Kingston, P. 2001. "The unfulfilled promise of cultural capital theory", *Sociology and Education* 74(extra issue): 88–99.

Kinnear, S., Kabir, Z., Mann, J., and Bricknell, L. 2013. "The need to measure and manage the cumulative impacts of resource development on public health: an Australian perspective". In A. Rodriguez-Morales (ed.), *Current topics in public health* (pp. 125–148). Rijeka, Croatia: InTech Publishers.

Kirchner, C. 1988. "Western Europe: Subsidized survival". In M. Peck (ed.), *The world aluminum industry in a changing energy era* (pp. 61–89). New York: Routledge.

Kondo, M., Rivera, R., and Rullman, S. 2012. "Protecting the idyll but not the environment: Second homes, amenity migration and rural exclusion in Washington State", *Landscape and Urban Planning* 106(2): 174–182.

Kotilainen, J., Halonen, M., Vatanen, E., and Tykkyläinen, M. 2017. "Resource town transitions in Finland: Local impacts and policy responses in Lieksa". In G. Halseth (ed.), *Transformation of resource towns and peripheries: Political economy perspectives* (pp. 296–316). Oxford and New York: Routledge.

Kourtit, K., Nijkamp, P., Lowik, S., van Vught, F., and Vulto, P. 2011. "From islands of innovation to creative hotspots", *Regional Science Policy and Practice* 3(3): 145–161.

Kusakabe, E. 2012. "Social capital networks for achieving sustainable development", *Local Environment* 17(10): 1043–1062.

Kwan, M. and Schwanen, T. 2016. "Geographies of mobility", *Annals of the American Association of Geographers* 106(2): 243–256.

Lahiri-Dutt, K. 2012. "Digging women: Towards a new agenda for feminist critiques of mining", *Gender, Place and Culture* 19(2): 193–212.

Lassus, L., Lopez, S., and Roscigno, V. 2015. "Aging workers and the experience of job loss", *Research in Social Stratification and Mobility* 41: 81–91.

Lee, R. 2002. "'Nice maps, shame about the theory'? Thinking geographically about the economic", *Progress in Human Geography* 26(3): 333–355.

Lennie, J. 2010. *Learnings, case studies and guidelines for establishing shared and collaborative service delivery in the non-government sector: Evaluation of the Multi-Tenant Service Centre MTSC Pilots Project.* Brisbane: Department of Communities, Queensland Government, Brisbane.

Levebvre, H. 1991. *The production of space.* Cambridge, UK: Blackwell Publishers.

Levernier, W., Partridge, M., and Rickman, D. 2000. "The causes of regional variations in U.S. poverty: A cross-country analysis", *Journal of Regional Science* 40(3): 473–497.

Ley, D. 1983. *A social geography of the city.* New York: Harper and Row.

Little, J. 2002. *Gender and rural geography: Identity, sexuality and power in the countryside.* London and New York: Pearson Education Ltd.

Livingstone, D. 1992. *The geographical tradition: Episodes in the history of a contested enterprise.* Oxford, UK: Wiley-Blackwell Publishers.

Luxton, M. 1980. *More than a labour of love.* Toronto: The Women's Press.

MacKinnon, D. 2013. "Strategic coupling and regional development in resource economies: The case of the Pilbara", *Australian Geographer* 44(3): 305–321.

Maclean, K., Robinson, C., and Natcher, D. 2015. "Consensus building or constructive conflict? Aboriginal discursive strategies to enhance participation in natural resource management in Australia and Canada", *Society and Natural Resources* 28(2): 197–211.

MacLeod, G. 2001. "Beyond soft institutionalism: Accumulation, regulation, and their geographical fixes", *Environment and Planning Series A* 33(7): 1145–1167.

Major, C. and Winters, T. 2013. "Community by necessity: Security, insecurity, and the flattening of class in Fort McMurray, Alberta", *Canadian Journal of Sociology* 38(2): 141–165.

Makkonen, T. and Inkinen, T. 2014. "Spatial scaling of regional strategic programmes in Finland: A qualitative study of clusters and innovation systems", *Norsk Geografisk Tidsskrift-Norwegian Journal of Geography* 68(4): 216–227.

Makuwira, J. 2007. "The politics of community capacity-building: Contestations, contradictions, tensions and ambivalences in the discourse in Indigenous communities in Australia", *The Australian Journal of Indigenous Education* 36(S1): 129–136.

Manson, D., Markey, S., Ryser, L., and Halseth, G. 2016. "Recession response: Cyclical problems and local solutions in northern British Columbia", *Tijdschrift voor economische en sociale geografie* 107(1): 100–114.

Marchak, P. 1973. "Women workers and white-collar unions", *Canadian Review of Sociology and Anthropology* 10(2): 134–146.

Marchak, P. 2011. *Green gold: The forest industry in British Columbia*. Vancouver: UBC Press.

Markey, S. and Heisler, K. 2011. "Getting a fair share: Regional development in rapid boom-bust rural setting", *Canadian Journal of Regional Science* 33(3): 49–62.

Markey, S., Halseth, G., and Manson, D. 2006. "The struggle to compete: From comparative to competitive advantage in northern British Columbia", *International Planning Studies* 11(1): 19–39.

Markey, S., Halseth, G., and Manson, D. 2008a. "Challenging the inevitability of rural decline: Advancing the policy of place in northern British Columbia", *Journal of Rural Studies* 24(4): 409–421.

Markey, S., Halseth, G., and Manson, D. 2008b. "Closing the implementation gap: A framework for incorporating the context of place in economic development planning", *Local Environment* 13(4): 337–351.

Markey, S., Connelly, S., and Roseland, M. 2010. "Back of the envelope: Pragmatic planning for sustainable rural community development", *Planning, Practice, and Research* 25(1): 1–23.

Markey, S., Halseth, G., and Manson, D. 2012. *Investing in place: Economic renewal in northern British Columbia*. Vancouver: UBC Press.

Markey, S., Ryser, L., and Halseth, G. 2015. "We're in this all together: Community impacts of long-distance labour commuting", *Rural Society* 24(2): 131–153.

Marshall, J. 1999. "Voluntary activity and the state: Commentary and review of the literature relating to the role and impact of government involvement in rural communities in Canada". In B. Reimer (ed.), *Voluntary organizations in rural Canada: Final Report* (pp. 1. 1–1. 36). Montreal: Canadian Rural Restructuring Foundation, Concordia University.

Martin, A. 2014. "Entrenched instability: The community implications of flexibility in British Columbia's northern interior", *Journal of Rural and Community Development* 8(3): 159–173.

Martin, R. 1994. "Economic theory and human geography". In D. Gregory, R. Martin, and G. Smith (eds.), *Human geography: Society, space, and social science* (pp. 21–53). Minneapolis: University of Minnesota Press.

Martin, R. 2000. "Institutional approaches in economic geography". In E. Sheppard and T. Barnes (eds.), *A companion to economic geography* (pp. 77–94). Oxford: Blackwell Publishers.

Martin, R. 2010. "Roepke lecture in economic geography: Rethinking regional path dependence: Beyond lock-in to evolution", *Economic Geography* 86(1): 1–27.

Martin, R. 2012. "Replacing path dependence: A response to the debate", *International Journal of Urban and Regional Research* 36(1): 179–192.

Massam, B. 1995. *An essay on civil society.* Vancouver: Community Economic Development Centre, Simon Fraser University.

Massey, D. 1984a. "Introduction: Geography matters". In D. Massey and J. Allen (eds.), *Geography matters! – A reader* (pp. 1–11). Cambridge: Cambridge University Press.

Massey, D. 1984b. *Spatial divisions of labour: Social structures and the geography of production.* London: Macmillan.

Maunder, P. 2012. *Coal and the coast*, Christchurch, New Zealand: Canterbury University Press.

Mayes, R. 2015. "A social licence to operate: Corporate social responsibility, local communities and the constitution of global production networks", *Global Networks* 15(s1): S109–S128.

McAllister, M., Fitzpatrick, P., and Fonseca, A. 2014. "Challenges of space and place for corporate 'citizens' and healthy mining communities: The case of Logan Lake, BC and Highland Valley Copper", *The Extractive Industries and Society* 1(2): 312–320.

McCann, L. 1998. "Interpreting Canada's heartland and hinterland". In L. McCann and A. Gunn (eds.), *Heartland and hinterland: A regional geography of Canada*, 3rd edition (pp. 1–41). Scarborough, ON: Prentice-Hall Canada Inc.

McDonald, J. 2016. "Globalization and the transformation of Aboriginal society: The Tsimshian encounter". In P. Bowles and G. Wilson (eds.), *Resource communities in a globalizing region: development, agency, and contestation in northern British Columbia* (pp. 79–108). Vancouver: UBC Press.

McDonald, P., Mayes, R., and Pini, B. 2012a. "A spatially-oriented approach to the impact of the Ravensthorpe nickel mine closure in remote Australia", *Journal of Industrial Relations* 54(1): 22–40.

McDonald, P., Mayes, R., and Pini, B. 2012b. "Mining work, family and community: A spatially-oriented approach to the impact of the Ravensthorpe nickel mine closure in remote Australia", *Journal of Industrial Relations* 54(1): 22–40.

McDowell, L. 2001. "Father and Ford revisited: Gender, class and employment change in the new millennium", *Transactions of the Institute of British Geographers* 26(4): 448–464.

McIntyre, S. 2009. *A concise history of Australia*, 3rd edition. Cambridge: Cambridge University Press.

Mertins-Kirkwood, H. 2014. *Labour mobility in Canada: Issues and policy recommendations.* Ottawa: Canadian Labour Congress. Available online at: http://canadianlabour.ca/sites/default/files/media/LabourMobility-Rpt-HMK-2014-10-14.pdf. Accessed November 10, 2016.

Michelini, J. 2013. "Small farmers and social capital in development projects: Lessons from failures in Argentina's rural periphery", *Journal of Rural Studies* 30: 99–109.

Milbourne, L., Macrae, S., and Maquire, M. 2003. "Collaborative solutions or new policy problems: Exploring multi-agency partnerships in education and health work", *Journal of Education Policy* 18(1): 19–35.

Ministry of Social Development. 2014. *Satellite sites to remain closed until further notice.* Auckland, New Zealand: Ministry of Social Development. Available online at: https://www.msd.govt.nz/about-msd-and-our-work/newsroom/media-releases/2014/satellite-sites-to-remain-closed-until-further-notice.html. Accessed October 12, 2015.

Mitchell, C. and De Waal, S. 2009. "Revisiting the model of creative destruction: St. Jacobs, Ontario, a decade later", *Journal of Rural Studies* 25(1): 156–167.

Mitchell, C. and O'Neill, K. 2016. "Tracing economic transition in the mine towns of northern Ontario: An application of the resource-dependency model", *The Canadian Geographer* 60(1): 91–106.

Mitchell, D. 1983. *WAC Bennett and the rise of British Columbia.* Vancouver: Douglas and MacIntyre.

Moseley, M., Parker, G., and Wragg, A. 2004. "Multi-service outlets in rural England: The co-location of disparate services", *Planning, Practice, & Research* 19(4):375–391.

Mouat, J. 1995. *Roaring days: Rossland and the history of mining in British Columbia.* Vancouver: University of British Columbia Press.

Nardone, G., Sisto, R., and Lopolito, A. 2010. "Social Capital in the LEADER Initiative: A methodological approach", *Journal of Rural Studies* 26(1): 63–72.

Nel, E. 2015. "Evolving regional and local economic development in New Zealand", *Local Economy* 30(1): 67–77.

Nelsen, J., Scoble, M., and Ostry, A. 2010. "Sustainable socio-economic development in mining communities: North-central British Columbia perspectives", *International Journal of Mining, Reclamation, and Environment* 24(2): 163–179.

Newby, H. 1987. *Country life: A social history of rural England.* London: Weidenfeld and Nicolson.

Newhook, J., Neis, B., Jackson, L., and Roseman, S. 2011. "Employment-related mobility and the health of workers, families, and communities: The Canadian context", *Labour* 67: 121–156.

Nicholas, C. and Welters, R. 2016. "Exploring determinants of the extent of long distance commuting in Australia: Accounting for space", *Australian Geographer* 47(1): 103–120.

NORDREGIO. 2011. *The new rural Europe: Towards rural cohesion policy.* Stockholm: NORDREGIO, ESPON, and University of the Highlands and Islands.

Norton, W. 2004. *Human geography*, 5th edition. Don Mills, Ontario: Oxford University Press.

Nozick, M. 1999. "Sustainable development begins at home: Community solutions to global problems". In J. Pierce and A. Dale (eds.), *Communities, development, and sustainability across Canada* (pp. 3–23). Vancouver: UBC Press.

NWS. 2016. *North West Shelf LNG.* Available online at: http://www.nwsssc.com/project/participants. Accessed August 30, 2016.

O'Faircheallaigh, C. 2013. "Extractive industries and Indigenous peoples: A changing dynamic?", *Journal of Rural Studies* 30: 20–30.

Office of the Premier. 2013, Feb. 12. *New British Columbia Prosperity Fund will ensure lasting benefits.* Victoria: Province of British Columbia. Available online at: https://news.gov.bc.ca/stories/new-british-columbia-prosperity-fund-will-ensure-lasting-benefits. Accessed February 17, 2016.

O'Hagan, S. and Cecil, B. 2007. "A macro-level approach to examining Canada's primary industry towns in a knowledge economy", *Journal of Rural and Community Development* 2(2): 18–43.

Oman, C. 1996. *The policy challenges of globalisation and regionalisation.* Paris: OECD Development Center, Policy Brief Number 11.

Oosterlynck, S. 2012. "Path dependence: A political economy perspective", *International Journal of Urban and Regional Research* 36(1): 158–165.

O'Shaughnessy, S. and Krogman, N. 2011. "Gender as contradiction: From dichotomies to diversity in natural resource extraction", *Journal of Rural Studies* 27(2): 134–143.

Paagman, A., Tate, M., Furtmueller, E., and de Bloom, J. 2015. "An integrative literature review and empirical validation of motives for introducing shared services in government organizations", *International Journal of Information Management* 35: 110–123.

Paasi, A. 2009. "The resurgence of the 'region' and 'regional identity': Theoretical perspectives and empirical observations on regional dynamics in Europe", *Review of International Studies* 35(1): 121–146.

Packer, J., Spence, R., and Beare, E. 2002. "Building community partnerships: An Australian case study of sustainable community-based rural programmes", *Community Development Journal* 37(4): 316–326.

Page, B. 1996. "Across the great divide: Agriculture and industrial geography", *Economic Geography* 72(4): 376–397.

Paniagua, A. 2016. "An individual rural geography", *The Professional Geographer* 68(3): 1–8.

Parr, J. 1990. *The gender of breadwinners: Women, men, and change in two industrial towns, 1880–1950*. Toronto: University of Toronto Press.

Paül, V. and Haslam McKenzie, F. 2015. "'About time the regions were recognised': Interpreting region-building in Western Australia", *Australian Geographer* 46(3): 363–388.

Peacock, A. 1985. "Tumbler Ridge: A new style resource town", *Priorities* 13: 6–11.

Peck, J. 2005. "Struggling with the creative class", *International Journal of Urban and Regional Affairs* 29(4): 740–770.

Peck, J. 2010. *Constructions of neoliberal reason*. Oxford: Oxford University Press.

Peck, J. 2013. "Polanyi in the Pilbara", *Australian Geographer* 44(3): 243–264.

Peddle, C. 2011. "*The impact of the 2008 recession on youth sport programs in a local community*". Unpublished masters thesis. Windsor, Ontario: University of Windsor.

Pedersen, S. 2015. "The right to traditional resources and development program". In T. Gjertsen and G. Halseth (eds.), *Sustainable development in the Circumpolar North – From Tana, Norway to Oktemtsy, Yakutia, Russia: The Gargia conferences for local and regional development 2004–14* (pp. 192–200). Prince George, BC: UNBC Community Development Institute Publications Series. Tromso, Norway: Septentrio Academic Publishing of the University Library at UiT The Arctic University of Norway, Septentrio Conference Series, Number 1, 2015.

Peet, R. 2007. *Geography of power: The making of global economic policy*. London: Zed Books.

Petronas. 2012. Available online at http://petronas.com. Accessed April 23, 2013.

Plummer, P. and Tonts, M. 2013. "Income growth and employment in the Pilbara: An evolutionary analysis, 1980–2010", *Australian Geographer* 44(3): 227–241.

Polèse, M. and Shearmur, R. 2006. "A Canadian case study with thoughts on local development strategies", *Papers in Regional Science* 85(1): 23–46.

Polèse, M. and Simard, M. 2012. *The resource curse and regional development: Does the Dutch Disease apply to local economies? Evidence from Canada*. Montreal: Institut national de la recherche scientifique INRS-Urbanisation, Culture et Société.

Poole, D., Ferguson, M., DiNitto, D., and Schwab, A. 2002. "The capacity of community-based organizations to lead local innovations in welfare reform: Early findings from Texas", *Nonprofit Management and Leadership* 12(3): 261–276.

Porter, E. 1999. "Women's deployment with husband's unemployment: Women's labour in the balance". In A. Mawhiney and J. Pitblado (eds.), *Boom town blues: Elliot Lake, collapse and revival in a single-industry community* (pp. 89–105). Toronto and Oxford: Dundurn Press.

Porter, M. 2004. *Competitiveness in rural US regions: Learning and research agenda.* Boston, MA: Institute for Strategy and Competitiveness, Harvard Business School.

Porter, M. and Kramer, M. 2011. "Creating shared value: How to reinvent new capitalism, and unleash a new wave of innovation and growth", *Harvard Business Review* January / February: 62–77.

Preston, V., Rose, D., Norcliffe, G., and Holmes, J. 2000. "Shift work, childcare and domestic work: Divisions of labour in Canadian paper mill communities", *Gender, Place and Culture* 7(1): 5–30.

Priaro, M. 2015. "Jim Prentice's government should revitalize Peter Lougheed's legacy". *Inside Policy – The Magazine of The Macdonald-Laurier Institute* February: 26–27.

Prudham, S. 2008. "Tall among the trees: Organizing against global forestry in rural British Columbia", *Journal of Rural Studies* 24(2): 182–196.

Putnam, R. 2000. *Bowling alone: The collapse and revival of American community.* New York: Simon and Schuster.

Ritzer, G. and Dean, P. 2015. *Globalization: A basic text*, 2nd edition. West Sussex, UK: Wiley Blackwell.

Reed, M. 1999. *The places of women in forestry communities on northern Vancouver Island.* FRBC Report. Burnaby, BC: FRBC, Research Programme, Communities Committee.

Reed, M. 2003a. "Marginality and gender at work in forestry communities of British Columbia, Canada", *Journal of Rural Studies* 19(3): 373–389.

Reed, M. 2003b. *Taking stands: Gender and the sustainability of rural communities.* Vancouver: UBC Press.

Reed, M. G. 2008. "Reproducing the gender order in Canadian forestry: The role of statistical representation", *Scandinavian Journal of Forest Research* 23(1): 78–91.

Rolfe, J. and Kinnear, S. 2013. "Populating regional Australia: What are the impacts of non-resident labour force practices on demographic growth in resource regions?", *Rural Society* 22(2): 125–137.

Roper, B. 2005. *Prosperity for all: Economic, social and political change in New Zealand since 1935.* Auckland, New Zealand: Thomson.

Royal Commission on Aboriginal Peoples. 1996. *People to people, nation to nation: Highlights from the Report of the Royal Commission on Aboriginal Peoples.* Ottawa: Indigenous and Northern Affairs Canada. Available online at: http://www.ainc-inac. gc.ca/ap/pubs/rpt/rpt-eng.asp. Accessed March 12, 2014.

Ryser, L. and Halseth, G. 2011a. "Housing costs in an oil and gas boom town: issues for low-income senior women living alone", *Journal of Housing for the Elderly* 25(3): 306–325.

Ryser, L. and Halseth, G. 2011b. "Informal support networks of low-income senior women living alone: Evidence from Fort St. John, BC", *Journal of Women and Aging* 23(3): 185–202.

Ryser, L. and Halseth, G. 2012. "Challenges for engaging social economy business in rural and small town renewal". In L. Mook, J. Quarter, and S. Ryan (eds.), *Businesses with a difference: Balancing the social and the economic* (pp. 161–182). Toronto: University of Toronto Press.

Ryser, L. and Halseth, G. 2013. *Tracking the social and economic transformation process in Kitimat, BC.* Prince George, BC: UNBC Community Development Institute.

Ryser, L. and Halseth, G. 2016. "Opportunities and challenges to address poverty in rural regions: A case study from northern BC", *Journal of Poverty.* Available online at: http://dx.doi.org/10.1080/10875549.2016.1141386. Accessed November 10, 2016.

Ryser, L., Rajput, A., Halseth, G., and Markey, S. 2012a. *Hollowing out the community: Community impacts of extended long distance labour commuting.* Prince George, BC: UNBC Community Development Institute. Available online at: http://www.unbc.ca/sites/default/files/assets/community_development_institute/research/maccommute/hollowing_out_the_community_summary_report_july_2012.pdf. Accessed March 2, 2015.

Ryser, L., Schwamborn, J., and Halseth, G. 2012b. *Lessons from economic upswings: A case study of the Peace River Region.* Prince George, BC: UNBC Community Development Institute. Available online at: http://www.unbc.ca/sites/default/files/assets/community_development_institute/research/ktids/ktids_summary_theme_report_copy1.pdf. Accessed March 2, 2015.

Ryser, L., Markey, S., Manson, D., Schwamborn, J., and Halseth, G. 2014. "From boom and bust to regional waves: Development patterns in the Peace River Region, British Columbia", *Journal of Rural and Community Development* 9(1): 87–111.

Ryser, L., Good, J., Morris, M., Halseth, G., and Markey, S. 2015. *Best practices guiding industry-community relationships, planning, and mobile workforces.* Prepared for the BC Natural Gas Workforce Strategy Committee. Prince George, BC: Community Development Institute, University of Northern British Columbia. Available online at: http://www.unbc.ca/sites/default/files/news/40513/lessons-learned-work-camp-community-relations-practices-making-positive-difference/best_practices_guiding_industry-community_relationships_and_mobile_workforces_final_-_march_2015.pdf. Accessed April 23, 2015.

Ryser, L., Patterson, D., Bodhi, M., Halseth, G., Good, J., Markey, S., and Naghshinepour Esfahani, N. 2016a. *Learning from smart services and infrastructure projects in rural BC: Final report.* Prepared for the Rural Policy Learning Commons. Prince George, BC: Canada Research Chair in Rural and Small Town Studies, University of Northern British Columbia.

Ryser, L., Markey, S., and Halseth, G. 2016b. "The workers' perspective: The impacts of long distance labour commuting in a northern Canadian small town", *The Extractive Industries and Society* 3(3): 594–605.

Sæþórsdóttir, A. and Saarinen, J. 2016. "Challenges due to changing ideas of natural resources: Tourism and power plant development in the Icelandic wilderness", *Polar Record* 52(1): 82–91.

Sampson, R. 2012. "What community supplies". In J. DeFilippis and S. Saegert (eds.), *The Community development reader,* 2nd edition (pp. 308–318). New York: Routledge.

Sanderson, D. and Polson, R. 1939. *Rural community organization.* New York: Wiley.

Sharma, S. 2010. "The impact of mining on women: Lessons from the coal mining Bowen Basin of Queensland, Australia", *Impact Assessment and Project Appraisal* 28(3): 201–215.

Shearmur, R. and Bonnet, N. 2011. "Does local technological innovation lead to local development? A policy perspective", *Regional Science Policy and Practice* 3(3): 249–270.

Sheppard, E. 2009. "Political economy". In D. Gregory, R. Johnston, G. Pratt, M. Watts, and S. Whatmore (eds.), *The dictionary of human geography*, 5th edition (pp. 547–549). Oxford: Wiley-Blackwell.

Sheppard, E. 2013. "Thinking through the Pilbara", *Australian Geographer* 44(3): 265–282.

Sheppard, E. 2015. "Thinking geographically: Globalizing capitalism and beyond", *Annals of the Association of American Geographers* 105(6): 1113–1134.

Sherval, M. and Hardiman, K. 2014. "Competing perceptions of the rural idyll: Responses to threats from coal seam gas development in Gloucester, NSW, Australia", *Australian Geographer* 45(2): 185–203.

Shortall, S. 2008. "Are rural development programmes socially inclusive? Social inclusion, civic engagement, participation, and social capital: Exploring the differences", *Journal of Rural Studies* 24(4): 450–457.

Shrimpton, M. and Storey, K. 1992. "Fly-in mining and the future of the Canadian North". In M. Bray and A. Thomson (eds.), *At the end of the shift: Mines and single-industry towns in northern Ontario* (pp. 187–208). Toronto: Dundurn Press Limited.

Shucksmith, M. 2004. "Young people and social exclusion in rural areas", *Sociologia Ruralis* 44(1): 43–59.

Shucksmith, M. 2009. "Disintegrated rural development? Neo-endogenous rural development, planning and place-shaping in diffused power contexts", *Sociologia Ruralis* 50(1): 1–14.

Simandan, D. 2012. "Rethinking the conceptual foundations of evolutionary economic geography: Introduction to a debate", *International Journal of Urban and Regional Research* 36(1): 156–157.

Simpson, G. and Clifton, J. 2010. "Funding and facilitation: Implications of changing government policy for the future of voluntary Landcare groups in Western Australia", *Australian Geographer* 41(3): 403–423.

Simpson, J. 2012. "Alberta's flushing its resource miracle down the drain". *The Globe and Mail* February 10. Available online at: http://www.theglobeandmail.com/op inion/albertas-flushing-its-resource-miracle-down-the-drain/article4420395. Accessed February 17, 2016.

Sinha, M. 2015. *Volunteering in Canada, 2004–2013*. Ottawa: Statistics Canada. Available online at: http://volunteeralberta.ab.ca/wp-content/uploads/2015/11/Volun teering-in-Canada-2004-2013.pdf. Accessed November 2, 2016.

Skinner, M. and Hanlon, N. 2016. "Introduction to ageing resource communities". In M. Skinner and N. Hanlon (eds.), *Ageing resource communities: New frontiers of rural population change, community development, and voluntarism* (pp. 1–8). London: Routledge.

Skinner, M. and Joseph, A. 2011. "Placing voluntarism within evolving spaces of care in ageing rural communities", *GeoJournal* 76(2): 151–162.

Skúlason, J. and Hayter, R. 1998. "Industrial location as a Bargain: Iceland and the aluminium multinationals 1962–1994", *Geografiska Annaler: Series B, Human Geography* 80(1): 29–48.

Smith, N. 1996. "Spaces of vulnerability: The space of flows and the politics of scale", *Critique of Anthropology* 16(1): 63–77.

Smyth, P., Reddel, T., and Jones, A. 2004. "Social inclusion, new regionalism, and associational governance: The Queensland experience", *International Journal of Urban and Regional Research* 38(3), 601–615.

Sorensen, T. 2017. "Community development in an age of mounting uncertainty: Armidale, Australia". In G. Halseth (ed.), *Transformation of resource towns and peripheries: Political economy perspectives* (pp. 249–267). Oxford and New York: Routledge.

Sparke, M. 2002. "Not a state, but more than a state of mind: Cascading Cascadias and the geoeconomics of cross-border regionalism". In M. Perkmann and N. Sum (eds.), *Globalization, regionalization and cross-border regions* (pp. 221–238). New York: Palgrave Macmillan.

Status of Women Canada. 2013. *Women and girls in rural, remote and northern communities: Key to Canada's economic prosperity.* Available online at: http://www. swc-cfc.gc.ca/initiatives/wnc-fcn/intro-eng.pdf. Accessed February 6, 2016.

Stirling, C. M., O'Meara, P., Pedler, D., Tourle, V., and Walker, J. H. 2007. "Engaging rural communities in health care through a paramedic expanded scope of practice", *Rural and Remote Health* 7(4): 1–9.

St.-Martin, I. 1981. "Women in Schefferville: Research notes". In J. Bradbury and J. Wolfe (eds.), *Perspectives on social and economic change in the iron-ore mining region of Quebec-Labrador* (pp. 141–151). Montreal: McGill University Centre for Northern Studies and Research.

Storey, K. 2009. "The evolution of commute work in Canada and Australia". In G. Dzida, F. Stammler, G. Eilmsteiner-Saxinger, M. Pavlova, T. Vakhrusheva, Z. Borlakova, and M. Nourieva (eds.), *Biography, shift-labour and socialization in a northern industrial city – the far north: Particularities of labour and human socialization* (pp. 23–32). Proceedings of the International Conference in Novy Urengoy, Russia, 4–6 December 2008. Available online at: http://articcentre.ulapland.fi/docs/NURbook_2ed_100421_final.pdf. Accessed August 28, 2014.

Storey, K. 2010. "Fly-in/fly-out: Implications for community sustainability", *Sustainability* 2: 1161–1181.

Streek, W. and Thelen, K. (eds.). 2005. *Beyond continuity: Institutional change in advanced political economies.* Oxford: Oxford University Press.

Stutz, F. and de Souza, A. 1998. *The world economy: Resources, location, trade, and development*, 3rd edition. Upper Saddle River, NJ: Prentice-Hall.

Sullivan, L. and Halseth, G. 2004. "Responses of volunteer groups in rural Canada to changing funding and service needs: Mackenzie and Tumbler Ridge, British Columbia". In G. Halseth and R. Halseth (eds.), *Building for success: Explorations of rural community and rural development* (pp. 337–362). Brandon, Manitoba: Rural Development Institute and Canadian Rural Revitalization Foundation.

Sullivan, L., Ryser, L., and Halseth, G. 2014. "Recognizing change, recognizing rural: The new rural economy and towards a new model of rural service", *The Journal of Rural and Community Development* 9(4): 219–245.

Suorsa, K. 2014. "The concept of 'region' in research on regional innovation systems", *Norsk Geografisk Tidsskrift-Norwegian Journal of Geography* 68(4): 207–215.

Swarts, J. 2013. *Constructing neoliberalism: Economic transformation in Anglo-American democracies.* Toronto: University of Toronto Press.

Swyngedouw, E. 1986. *The socio-spatial implications of innovations in the industrial organization.* Working Paper No. 20. Johns Hopkins European Center for Regional Planning and Research.

Tencer, D. 2014. "Norway's oil fund heads for $1 trillion; So where is Alberta's pot of gold?" *Huffington Post Canada*. Available online at: http://www.huffingtonpost.ca/2014/01/11/oil-fund-norway-millionaires_n_4576887.html. Accessed February 2, 2016.

Tennberg, M., Vola, J., Espiritu, A., Fors, B., Ejdemo, T., Riabova, L., Korchak, E., Tonkova, E., and Nosova, T. 2014. "Neoliberal governance, sustainable development and local communities in the Barents Region", *Peoples, Economies and Politics* 1(1): 41–72.

Tonts, M. 2010. "Labour market dynamics in resource dependent regions: An examination of the Western Australian Goldfields", *Geographical Research* 48(2): 148–165.

Tonts, M. and Haslam-McKenzie, F. 2005. "Neoliberalism and changing regional policy in Australia", *International Planning Studies* 10(3–4): 183–200.

Tonts, M., Plummer, P., and Lawrie, M. 2012. "Socio-economic wellbeing in Australian mining towns: A comparative analysis", *Journal of Rural Studies* 28(3): 288–301.

Tonts, M., Martinus, K., and Plummer, P. 2013. "Regional development, redistribution and the extraction of mineral resources: the Western Australian Goldfields as a resource bank", *Applied Geography* 45: 365–374.

Torkington, A., Larkins, S., and Gupta, T. 2011. "The psychosocial impacts of fly-in fly-out and drive-in drive-out mining on mining employees: A qualitative study", *Australian Journal of Rural Health* 19(3): 135–141.

Trigger, D., Keenan, J., de Rijke, K., and Rifkin, W. 2014. "Aboriginal engagement and agreement-making with a rapidly developing resource industry: Coal seam gas development in Australia", *The Extractive Industries and Society* 1(2): 176–188.

Troughton, M. 2005. "Fordism rampant: The model and reality, as applied to production, processing and distribution in the North American agro-food system". In S. Essex, A. Gilg, R. Yarwood, J. Smithers, and R. Wilson (eds.), *Rural change and sustainability: Agriculture, the environment and communities* (pp. 13–27). Wallingford, Oxfordshire, UK: CAB International.

Truth and Reconciliation Commission of Canada. 2015. *Honouring the truth, reconciling for the future: Summary of the final report of the Truth and Reconciliation Commission of Canada*. Winnipeg, Manitoba: Truth and Reconciliation Commission of Canada.

Tykkyläinen, M. 2006. "Dynamics of job creation, restructuring and industrialisation in rural Finland", *Fennia-International Journal of Geography* 184(2): 151–167.

Tykkyläinen, M., Vatanen, E., Halonen, M., and Kotilainen, J. 2017. "Global–local links and industrial restructuring in a resource town in Finland: The case of Lieksa". In G. Halseth (ed.), *Transformation of resource towns and peripheries: Political economy perspectives* (pp. 85–111). Oxford and New York: Routledge.

Uhlmann, V., Rifkin, W., Everingham, J., Head, B., and May, K. 2014. "Prioritising indicators of cumulative socio-economic impacts to characterise rapid development of onshore gas resources", *The Extractive Industries and Society* 1(2): 189–199.

Valentine, G. 2001. *Social geographies: Space and society*. Essex, England: Pearson Education Ltd.

Vodden, K. and Hall, H. 2016. "Long distance commuting in the mining and oil and gas sectors: Implications for rural regions", *The Extractive Industries and Society* 3(3): 577–583.

Waeyenberge, E., Fine, B., and Bayliss, K. 2011. "The World Bank, Neo-Liberalism and Development Research". In K. Bayless, B. Fine, and E. V. Waeyenberge (eds.),

The political economy of development: The world bank, neoliberalism and development research (pp. 3–25). New York and London: PlutoPress.

Walk, M., Schinnenburg, H., and Handy, F. 2013. "Missing in action: Strategic human resource management in German nonprofits", *Voluntas* Online First: 1–31.

Walsh, P. 2008. "Shared services: Lessons from the public and private sectors for the nonprofit sector", *The Australian Journal of Public Administration* 67(2): 200–212.

Warrack, A. and Keddie, R. 2002. *Natural resource trust funds: A comparison of Alberta and Alaska resource funds.* Edmonton, Alberta: Western Centre for Economic Research, University of Alberta, No. 72.

Watts, M. 2009. "Capital". In D. Gregory, R. Johnston, G. Pratt, M. Watts, and S. Whatmore (eds.), *The dictionary of human geography,* 5th edition (pp. 58–59). Oxford: Wiley-Blackwell.

Weaver, R. 2014. "Compliance regimes and barriers to behavioral change", *Governance* 27(2): 243–265.

West Fraser Timber. 2007. *Annual report.* Vancouver: West Fraser Timber.

Wetzstein, S. 2011. *Managing boomtown Perth: Policy changes for adequate housing provision.* FACTBase Bulletin 23. Perth, Australia: Committee for Perth, University of Western Australia.

Wiles, J. and Jayasinha, R. 2013. "Care for place: The contributions older people make to their communities", *Journal of Aging Studies* 27(2): 93–101.

Williston, E. and Keller, B. 1997. *Forests, power, and policy: The legacy of Ray Williston.* Prince George, BC: Caitlin Press.

Wilson, L. 2004. "Riding the resource roller coaster: Understanding socioeconomic differences between mining communities", *Rural Sociology* 69(2): 261–281.

Wollebæk, D. 2009. "Survival in local voluntary association", *Nonprofit Management and Leadership* 19(3): 267–284.

Wollebæk, D. and Selle, P. 2004. "The role of women in the transformation of the organizational society in Norway", *Nonprofit and Voluntary Sector Quarterly* 33(3): 120S–144S.

Woods, M. 2007. "Engaging the global countryside: Globalization, hybridity and the reconstitution of rural place", *Progress in Human Geography* 31(4): 485–507.

Woods, M. 2010. "The political economies of place in the emergent global countryside: Stories from rural Wales". In G. Halseth, S. Markey, and D. Bruce (eds.), *The Next Rural Economies: Constructing rural place in global economies* (pp. 166–178). Wallingford, Oxfordshire, UK: CAB International.

Woodside Energy Ltd. 2016. *Pluto LNG.* Available online at: http://www.woodside. com.au/Our-Business/Producing/Pages/Pluto.aspx#.V8YPvzWBt40. Accessed August 30, 2016.

Woolvin, M., Atterton, J., and Skerratt, S. 2012. *Rural parliaments in Europe: A report for the Scottish Government.* Edinburgh: Rural Policy Centre.

Wray, D. 2012. "'Daddy Lives at the Airport': The Consequences of Economically Driven Separation on Family Life in the Post-Industrial Mining Communities of Cape Breton", *Employee Responsibilities and Rights Journal* 24(2): 147–158.

Young, N. and Matthews, R. 2007. "Resource economies and neoliberal experimentation: The reform of industry and community in rural British Columbia", *Area* 39(2): 176–185.

Index

Printed and bound by CPI Group (UK) Ltd, Croydon, CR0 4YY

23/10/2024

01778241-0019